Introductory Eng

Under the Editorship

BSc(Eng), PhD, DIC, ACGI, MIMechE

Foundations of engineering mechanics

Professor G.R. Higginson,
BSc, PhD, MICE, MI MechE,
Department of Engineering Science,
University of Durham

Longman

Longman
1724-1974

LONGMAN GROUP LIMITED
London

Associated companies, branches and representatives
throughout the world

First published 1974

ISBN 0 582 44307.5 cased
ISBN 0 582 44279.6 paper

Library of Congress Catalog Card Number 93-88375

PRINTED BY Unwin Brothers Limited
THE GRESHAM PRESS OLD WOKING SURREY ENGLAND

Produced by 'Uneoprint'

A member of the Staples Printing Group

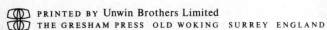

Preface

In my experience prefaces are not read unless they are very short. This book is written for engineering undergraduates, not teachers, but it would be gratifying if it were of some interest to both. Its aim is to create interest, rather than to convey a lot of information.

I would like to thank Dr G. A. Webster and Professor D. E. Newland for reading the manuscript and making valuable suggestions; and Mr J. Moseley for drawing the diagrams; and especially Miss Carole Anderson for typing the whole book.

G. R. Higginson

Durham, 1973

Acknowledgements

We are grateful for permission to reproduce the following illustration:
FIG. 7.24 from *The Aerodynamics of Boomerangs*, by Felix Hess,

Contents

1 Introduction

The purpose of this short introduction is to describe the role of mechanics in engineering as a whole. The most important function of the engineer is to make new things like cathedrals and aeroplanes for the benefit of other people, which is a creative function. But he has a responsibility to society not to make disastrous mistakes, so he must master the analytical side of his subject; then he can find out how existing things work and, more important, make a reasonable estimate of how, or whether, a new idea can be made to work.

Some of the masterpieces of engineering have been created by men who covered the whole range of disciplines: architecture, civil and mechanical engineering, and management. Men like I. K. Brunel and John Smeaton were giants over the whole field of their day. It is unlikely that one man now could be an authority in all aspects of engineering, but his education must give him a grasp of the fundamentals over the whole field if he is to understand the significance of his own contribution, and appreciate the importance of engineering to society as a whole.

An analytical approach to the main branches of engineering is based on a study of mathematics, electricity, the behaviour of solids and the behaviour of fluids. Of course these knit together in varying proportions to form the different specialities.

The engineer's interest in fluid mechanics developed from hydraulics, which is concerned with water movement, through hydrodynamics and aerodynamics to the modern fluid mechanics, which is so closely linked with thermodynamics as to be virtually the same subject. Thermodynamics grew out of the subject known in the first half of this century as Heat Engines. It has spread its wings far beyond reciprocating engines and allied devices.

The analysis of the behaviour of solids, to which this book is an introduction, is the subject known as solid mechanics or simply Engineering Mechanics.

The unity of all these subjects is found in the field of highest creative endeavour—design. A new design is usually the idea of one man; he expresses his idea in general terms, as an overall picture without details. The small detail is put in, examined, analysed, modified, and the whole built by a team who may individually specialise in one or other of the sub-divisions. But the man with the original idea, or anyone who is to master the whole concept, must have a grasp of the fundamentals of all the subjects. For example, the design of an internal combustion engine would involve thermodynamics, dynamics, stress analysis, properties of materials and many mathematical or computational techniques, plus the sort of judgement that only experience brings. Now think about designing an aeroplane !

The object of this book, which is one of a series of introductory texts for engineers, is to bring into focus those fundamentals in mechanics which are of particular importance in studying the behaviour of solids in engineering applications. Not all the examples are of real or even text-book engineering situations; those are largely left, in detail at any rate, to the more advanced texts. The author of an introductory text has the pleasant freedom to dodge back and forth between the real world of engines and collapsing blocks of flats, and the mathematical retreat where power is transmitted by light inextensible strings passing over frictionless pulleys, and a favourite pastime is rolling small spheres on big spheres.

After a brief look at some jargon and tools of the trade in the next chapter, Newton's laws are recalled in Chapter 3. The remaining chapters lead the way to engineering applications of statics and dynamics.

2 Background

This chapter contains the sort of background information that will be familiar to many readers, but not to all: first a few engineering terms and an outline of what constitutes engineering mechanics, then an introduction to vectors as a tool in mechanics, with just enough vector algebra to get us through the later chapters.

A simple and commonly used division of real engineering mechanics is into dynamics and statics—i.e. into things that 'go' and things that don't 'go'. Roughly speaking, things that go are called *machines* and things that don't go are called *structures*.

A structure is an assembly of bodies connected together in such a way as to become in effect a single rigid body, capable of transmitting forces without significant change of shape. Examples are shown in Fig. 2.1.

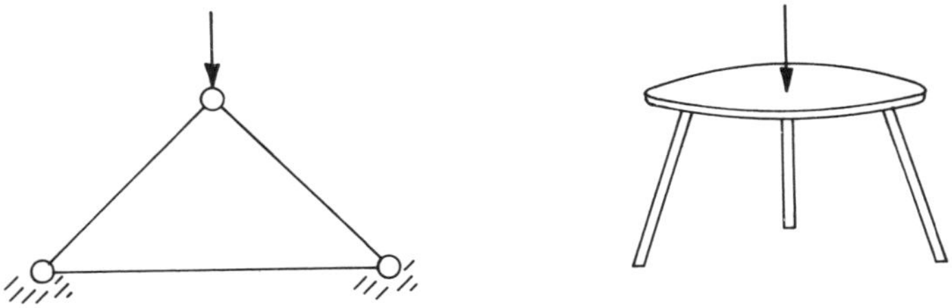

FIG. 2.1 Simple structures

A machine on the other hand is an assembly of bodies with freedom of motion, but only with sufficient freedom to move in an ordered way. In either of the machines in Fig. 2. 2, if the position of the link AB is given, the positions of all the other components are determined.

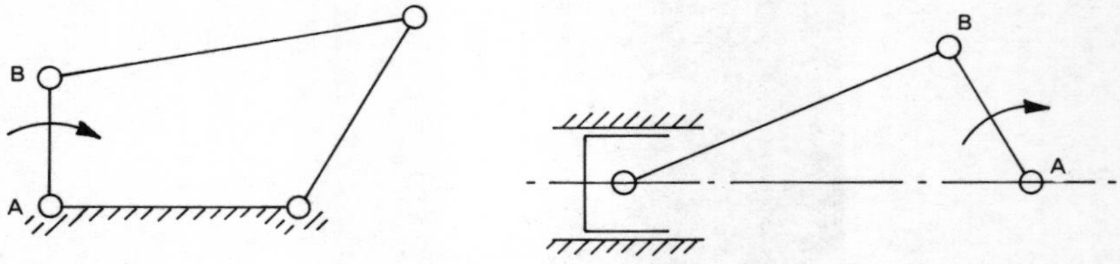

FIG. 2. 2

A machine is capable of transmitting motion and forces; in other words it can transmit *power*. The special case of a device which transmits motion only, with negligible forces, is usually called a *mechanism*.

A sort of half-way house between structures and machines is a flexible structure or mechanism which is capable of substantial but limited motion, and in which the flexibility is vested chiefly in one member. An example is a vehicle suspension with a spring as the flexible member, of the kind shown in Fig. 2. 3.

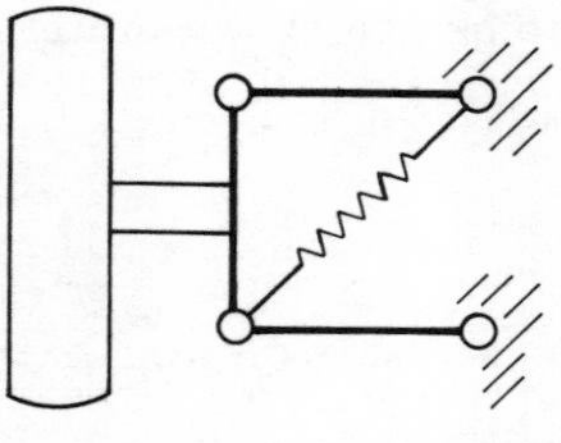

FIG. 2. 3 Vehicle suspension

There is clearly a critical number of members in a structure, as opposed to a machine. Assembled in a plane, for example, with 'pin joints', four members as in Fig. 2. 4 (a) form a mechanism or machine, while a fifth member as in (b) makes the collection into a structure. Figure 2. 4 (a) shows the basis of all plane mechanisms or machines, the so-called *'four-bar chain'*. Figure 2. 4 (b)

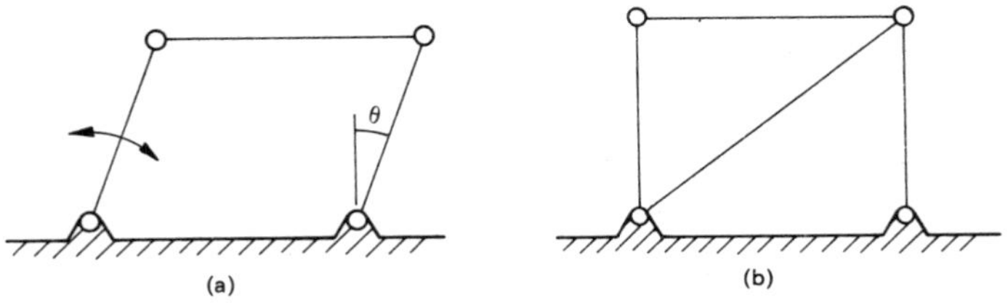

FIG. 2. 4

on the other hand shows plainly that the *triangle* is the basis of the simple plane structure; each additional intersection (or joint) must be located by two rigid members which are in turn pinned at fixed points. Similarly, to locate each additional point in a three-dimensional 'space-frame' three members are required.

If the number of members is one fewer than that required to make a plane frame, then the assembly is a mechanism. But the motion of a mechanism is considerably constrained; in fact there is only one way in which it can move—it is said to have only one 'degree of freedom'.

Degrees of freedom

The position of the mechanism shown in Fig. 2. 4 (a) can be described completely by one quantity, for example the angle θ. That is why it is said to have one degree of freedom. Similarly the position of a train on a track out of King's Cross can be defined by one quantity-distance. On the other hand a motor car in a car park needs three such quantities, two to define where it is and the third to give the direction in which it is pointing. If we consider the motion of an aeorplane in flight it is clear that three coordinates are required to locate a point on the aeroplane in the sky (its centre of mass perhaps); and that to define its direction three angles are needed. In other words, it has six degrees of freedom, and that is the general property of a rigid body free to move in three dimensions.

Vectors

The development of the algebra and arithmetic of vector quantities has simplified the analytical side of mechanics very much, so we must clearly take advantage of it.

Anyone with any experience of force or velocity, or the many other physical quantities which have *direction* as well as *magnitude*, already has a feel for vectors and the way they must be added together. Such a feel will alone suffice for almost all of this book, but there are a few points where the products of vectors are used, and so it is appropriate that they should be introduced in this chapter. But first we remind ourselves of some of the simple properties of vectors.

A vector is an entity characterised by *magnitude* and *direction*. In contrast a *scalar* is an entity with magnitude but no directional property. A vector may be a 'free vector' or a 'localised vector'. In Fig. 2. 5 the lines AB, CD, EF all have the same length and direction. If they represent, for example, the wind velocity, their location is immaterial, so they are all the same free vector. In the case of a force the point of application is important, so a force is a localised vector.

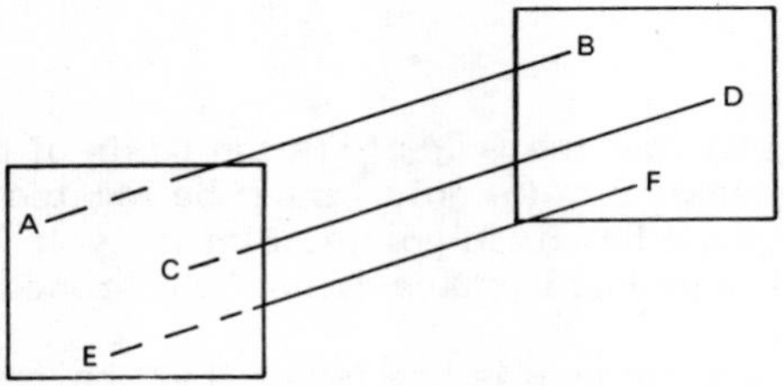

FIG. 2. 5

Free vectors or localised vectors that intersect can be added or subtracted directly by the parallelogram law. The vectors **b** and **c*** represented by

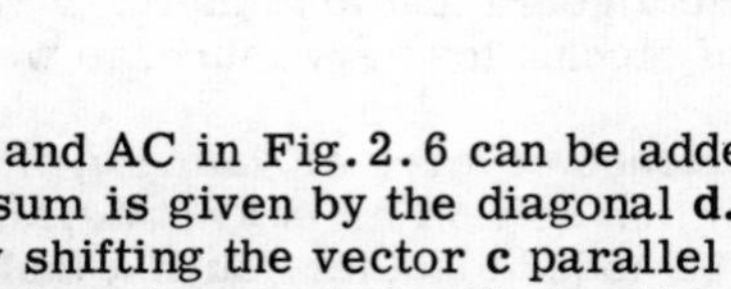

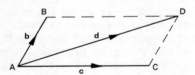

FIG. 2. 6 Parallelogram of vectors

the lines AB and AC in Fig. 2. 6 can be added by completing the parallelogram ABDC. The sum is given by the diagonal **d**. The construction is sometimes simplified by shifting the vector **c** parallel to itself to the tip B of vector **b** and drawing only the triangle ABD. It should be noted that in such a construction the

* It is common practice to represent vectors by letters in bold **type.** In handwriting this is usually replaced by a wavy line under the letter; that is the printer's symbol for bold type.

direction of the vector **c** (given by the arrow) must follow on the direction of the vector **b**; see Fig. 2.7. The sum **d** of the vectors **b** and **c** is called the *resultant* of **b** and **c**.

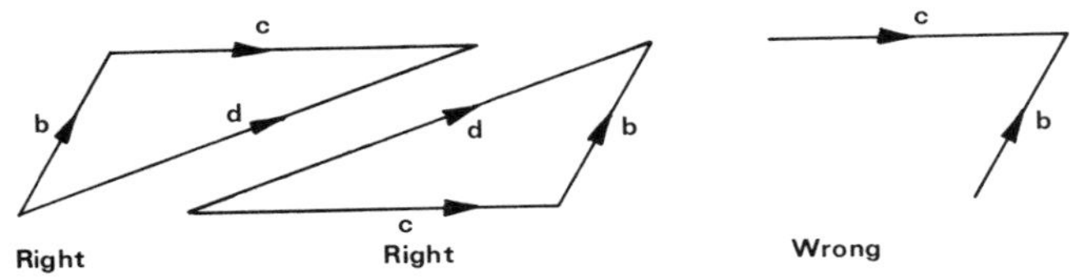

FIG. 2. 7 Triangle of vectors

It is easy to see, aided if necessary by a bit of sketching, that any number of free or intersecting localised vectors can be added by extending the triangle of vectors into a 'polygon of vectors' as shown in Fig. 2. 8. It is also

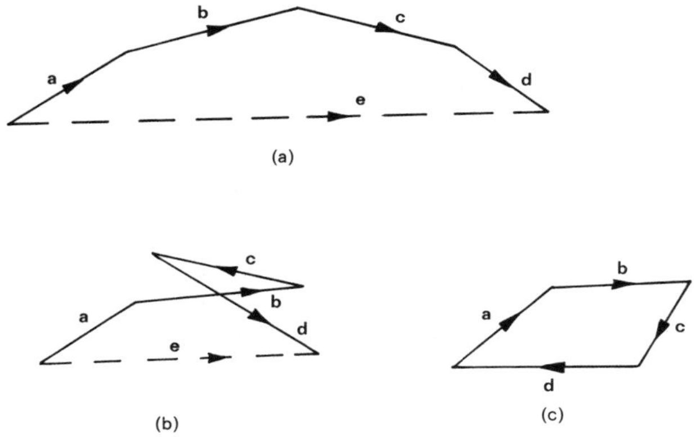

FIG. 2. 8 Vector polygons

clear that these vectors need not be in a plane, but can just as well be in three dimensions. In (a) and (b) the vectors **a, b, c** and **d** have as their resultant **e**. If the vectors **a, b, c** and **d** form a closed polygon, as in Fig. 2. 8 (c), then their sum, or resultant, is zero.

As an example of this last point, consider three forces acting on a body as in Fig. 2. 9. If the forces intersect as in Fig. 2. 9 (a) it can be seen that

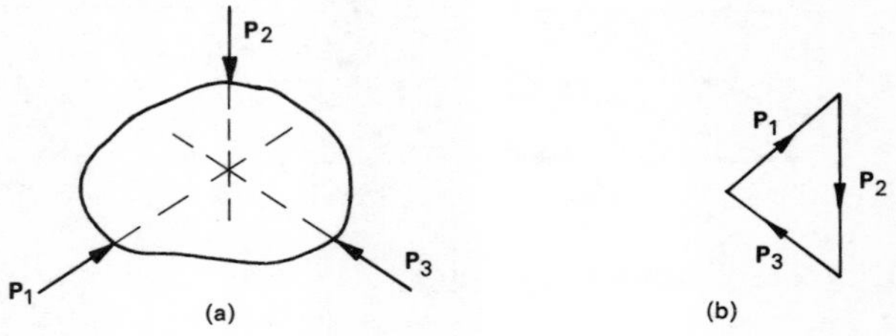

FIG. 2. 9 Three forces in equilibrium

they will be in *equilibrium* (i.e. they will have a zero resultant force) if the vector diagram formed by the three forces is a closed triangle as in Fig. 2. 9 (b). Clearly this is only possible if the three forces lie in the same plane. This is an important result and is worth repeating.

For a system of three forces to be in equilibrium they must all lie in the same plane and must intersect in a single point. They must also form a closed triangle of forces.

The same applies to any set of three localised vectors.

It should be noted that finite rotation is not a vector, even though it has magnitude and direction. That is because the resultant of a number of finite rotations is not independent of the order in which they occur. (You can easily check for yourself by taking a solid object like a match-box and setting it in a pair of x-y axes: (a) rotate it through a right-angle clockwise about the x axis, then about the y axis; (b) go back to the original position and rotate first about the y axis and then the x axis; compare the final positions.) *Infinitesimal* rotation, and therefore angular velocity and acceleration, *are* vectors.

Vector algebra

Only a few of the simpler statements of vector algebra will be given here. For a fuller treatment see Rutherford [1]* or Simons [2].

* Figures in square brackets indicate references at end of the chapters.

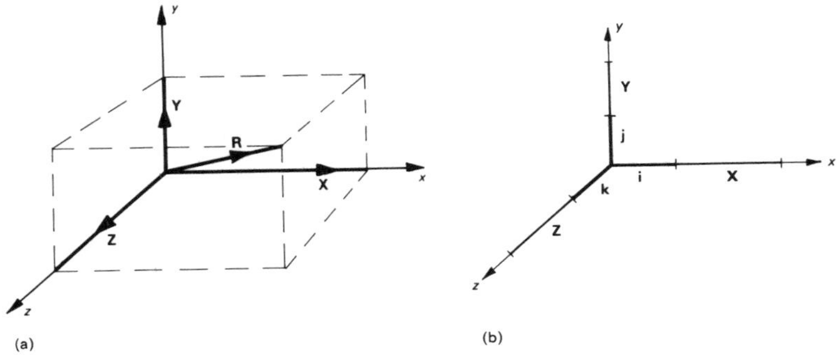

FIG. 2.10 Unit Vectors

A convenient representation of a vector, very often, is in terms of *unit vectors*, associated with a particular set of axes. Consider the vector **R** shown in the (xyz) frame in Fig. 2.10. First 'decompose' or 'resolve' **R** into the components **X, Y, Z** (the opposite of adding by the vector polygon) as in Fig. 2.10 (a). If we now define **i, j** and **k** as *unit vectors* in the x, y and z directions, we can describe **X, Y, Z** as

$$\mathbf{X} = x\mathbf{i}$$

$$\mathbf{Y} = y\mathbf{j}$$

$$\mathbf{Z} = z\mathbf{k}$$

The vectors **i, j, k** have *unit* magnitude in the directions of the x, y, z axes, while x, y and z simply denote the *magnitudes* of **X, Y** and **Z**. So $\mathbf{X} = x\mathbf{i}$ describes a vector of direction **i** and magnitude x. And

$$\mathbf{R} = x\mathbf{i} + y\mathbf{j} + z\mathbf{k}$$

Clearly the magnitude of **R**, sometimes signified by $|\mathbf{R}|$ or often just by R, is

$$R = |\mathbf{R}| = (x^2 + y^2 + z^2)^{1/2}$$

Multiplication of a vector **P** by a scalar m, simply defines a second vector in the same direction as **P** but with different magnitude, $m\mathbf{P}$.

Products of two vectors

Two quite distinct products of two vectors are defined: they are the *scalar product* and the *vector product*. The word 'defined' should be emphasised; these products are simply defined as part of the basis of vector algebra.

The *scalar product* of two vectors **A** and **B** is written as **A . B** and is defined as a *scalar* of magnitude $AB \cos \theta$, where A and B are the magnitudes of **A** and **B** and θ is the angle between them. If the two vectors do not intersect, but are 'skew', θ is found by shifting one of the vectors parallel to itself until they do intersect. The product is usually described in words as 'A dot B'.

$$\mathbf{A} . \mathbf{B} = AB \cos \theta \quad \text{a scalar,}$$

where $A = |\mathbf{A}|$ and $B = |\mathbf{B}|$

Since the product is a scalar with no direction, but only magnitude, it follows that

$$\mathbf{A} . \mathbf{B} = \mathbf{B} . \mathbf{A}$$

If **A** and **B** have the same direction, $\cos \theta = 1$, and $\mathbf{A} . \mathbf{B} = AB$. In particular, $\mathbf{A} . \mathbf{A} = A^2$, examples being

$$\mathbf{i} . \mathbf{i} = \mathbf{j} . \mathbf{j} = \mathbf{k} . \mathbf{k} = 1$$

If **A** and **B** are perpendicular, $\cos \theta = 0$ and $\mathbf{A} . \mathbf{B} = 0$. So

$$\mathbf{i} . \mathbf{j} = \mathbf{j} . \mathbf{k} = \mathbf{k} . \mathbf{i} = 0$$

Scalar products involving a number of vectors go much as ordinary algebra, for example:

if $\mathbf{B} \neq \mathbf{C}_1 + \mathbf{C}_2 + \mathbf{C}_3$

then $\mathbf{A} . \mathbf{B} = \mathbf{A} . (\mathbf{C}_1 + \mathbf{C}_2 + \mathbf{C}_3) = \mathbf{A} . \mathbf{C}_1 + \mathbf{A} . \mathbf{C}_2 + \mathbf{A} . \mathbf{C}_3$

It follows from the last few lines that if

$$\mathbf{A} = a_x\mathbf{i} + a_y\mathbf{j} + a_z\mathbf{k}$$

and

$$\mathbf{B} = b_x\mathbf{i} + b_y\mathbf{j} + b_z\mathbf{k}$$

where a_x, b_x, etc., are magnitudes

then

$$\mathbf{A} . \mathbf{B} = a_x b_x + a_y b_y + a_z b_z \tag{2.1}$$

You should check this for yourself by multiplying out and using $\mathbf{i} . \mathbf{i} = 1, \mathbf{i} . \mathbf{j} = 0$, etc.

The *vector product* of two vectors **A** and **B** is written as $\mathbf{A} \wedge \mathbf{B}$ (or very often as $\mathbf{A} \times \mathbf{B}$) and is defined as a *vector* whose magnitude is $AB \sin \theta$, and whose direction is perpendicular to both **A** and **B** in a sense given by the 'right-hand screw rule', which is described below. In words the product is often called 'A vector B'.

The *right-hand screw rule* is illustrated in Fig. 2.11 which shows a

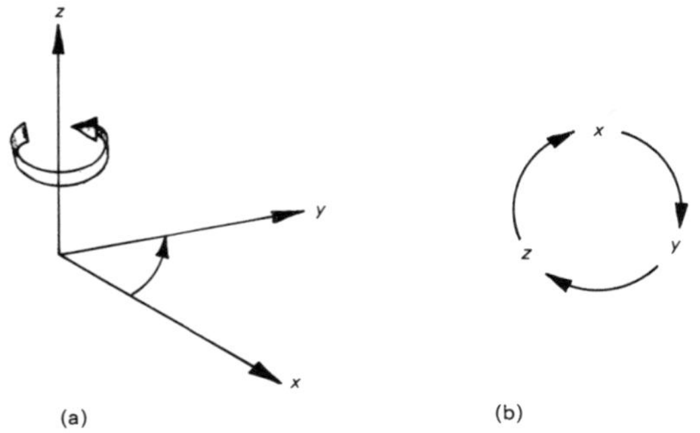

FIG. 2.11 Right-hand axes

set of 'right-hand axes'. The clockwise rotation from x to y by the shorter route (i.e. through the one right angle rather than the three) will advance a right-hand screw along the z-axis in the positive direction. Similarly $y \to z$ will advance a screw along the positive x direction. So the axes shown are right-handed in all the directions given by Fig. 2.11 (b).

Applying the right-hand screw rule now to the vector product, we see in Fig. 2.12 that $\mathbf{A} \wedge \mathbf{B}$ is directed upwards from the plane containing **A** and **B**. It

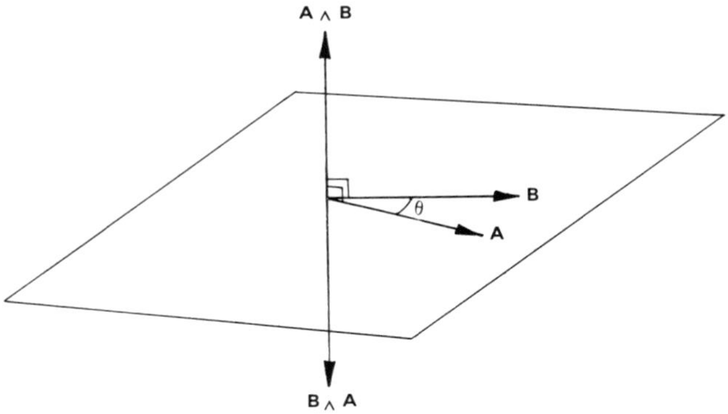

FIG. 2.12 Vector product

is quite plain however that $\mathbf{B} \wedge \mathbf{A}$ is directed downwards, so although the same in magnitude as $\mathbf{A} \wedge \mathbf{B}$, it is opposite in direction. Thus

$$\mathbf{B} \wedge \mathbf{A} = -\mathbf{A} \wedge \mathbf{B}$$

If $\mathbf{A}$ and $\mathbf{B}$ are parallel, $\sin\theta = 0$ and $\mathbf{A} \wedge \mathbf{B} = 0$. In particular

$$\mathbf{i} \wedge \mathbf{i} = \mathbf{j} \wedge \mathbf{j} = \mathbf{k} \wedge \mathbf{k} = 0$$

If $\mathbf{A}$ and $\mathbf{B}$ are perpendicular, $\sin\theta = 1$, or -1, the magnitude of $\mathbf{A} \wedge \mathbf{B}$ is AB. In particular,

$$\mathbf{i} \wedge \mathbf{j} = \mathbf{k}, \mathbf{j} \wedge \mathbf{k} = \mathbf{i}, \mathbf{k} \wedge \mathbf{i} = \mathbf{j}$$

and $$\mathbf{j} \wedge \mathbf{i} = -\mathbf{k}, \mathbf{k} \wedge \mathbf{j} = -\mathbf{i}, \mathbf{i} \wedge \mathbf{k} = -\mathbf{j}$$

It can be shown that

$$\mathbf{A} \wedge (\mathbf{B} + \mathbf{C}) = \mathbf{A} \wedge \mathbf{B} + \mathbf{A} \wedge \mathbf{C} \qquad (2.2$$

Vector products can therefore be expanded in the same way as products in ordinar algebra, but care must be taken to keep to the correct order of the factors. For example

$$(\mathbf{A} + \mathbf{B}) \wedge (\mathbf{C} + \mathbf{D}) = \mathbf{A} \wedge \mathbf{C} + \mathbf{A} \wedge \mathbf{D} + \mathbf{B} \wedge \mathbf{C} + \mathbf{B} \wedge \mathbf{D}$$

The vector product can now be simply expressed in unit vector and cartesian components as follows:

if $$\mathbf{A} = a_x\mathbf{i} + a_y\mathbf{j} + a_z\mathbf{k}$$

and $$\mathbf{B} = b_x\mathbf{i} + b_y\mathbf{j} + b_z\mathbf{k}$$

you should now check that the only terms remaining when the product $\mathbf{A} \wedge \mathbf{B}$ is expanded are

$$\mathbf{A} \wedge \mathbf{B} = \mathbf{i}\,(a_y b_z - a_z b_y) + \mathbf{j}\,(a_z b_x - a_x b_z) + \mathbf{k}\,(a_x b_y - a_y b_x) \qquad (2.3$$

If you have met determinants you will see that this can be written as the determinant

$$\mathbf{A} \wedge \mathbf{B} = \begin{vmatrix} \mathbf{i} & \mathbf{j} & \mathbf{k} \\ a_x & a_y & a_z \\ b_x & b_y & b_z \end{vmatrix} \qquad (2.4$$

There will be just one application later in the book of the triple vecto product, $\mathbf{A} \wedge (\mathbf{B} \wedge \mathbf{C})$. That is the vector product of the vectors $\mathbf{A}$ and $\mathbf{B} \wedge \mathbf{C}$. It is a vector which is perpendicular to $\mathbf{A}$ and, because it is also perpendicular to $\mathbf{B} \wedge \mathbf{C}$, it must lie in the plane containing $\mathbf{B}$ and $\mathbf{C}$. It can be shown by using equation (2.3) that

$$\mathbf{A} \wedge (\mathbf{B} \wedge \mathbf{C}) = (\mathbf{A} \cdot \mathbf{C})\mathbf{B} - (\mathbf{A} \cdot \mathbf{B})\mathbf{C} \qquad (2.5$$

Note that the terms in the brackets on the right-hand side are scalars, and so the expression is of the form $m\mathbf{B} - n\mathbf{C}$, which conforms with the fact that $\mathbf{A} \wedge (\mathbf{B} \wedge \mathbf{C})$ is a vector lying in the plane containing **B** and **C**.

Also it can be seen that

$$\mathbf{A} \wedge (\mathbf{B} \wedge \mathbf{C}) = -(\mathbf{B} \wedge \mathbf{C}) \wedge \mathbf{A} = (\mathbf{C} \wedge \mathbf{B}) \wedge \mathbf{A}$$

The position of the brackets is important, as plainly $\mathbf{A} \wedge (\mathbf{B} \wedge \mathbf{C}) \neq (\mathbf{A} \wedge \mathbf{B}) \wedge \mathbf{C}$ the first being a vector in the plane containing **B** and **C**, the second a vector in the plane containing **A** and **B**.

Differentiation of vectors

The meaning of differentiation of a vector will become clear as examples arise, but it is worth noting here that, because a vector has direction and magnitude, it can have a differential (or 'derivative') by virtue of variation of its *magnitude or its direction.* The formal process of differentiation of vectors and vector products follows the same lines as ordinary calculus; proofs can be found in the books referred to earlier, so only a few important results will be quoted here.

If **A**, **B** and **C** are functions of a scalar u,

$$\frac{d}{du}(\mathbf{A} + \mathbf{B}) = \frac{d\mathbf{A}}{du} + \frac{d\mathbf{B}}{du}$$

$$\frac{d}{du}(\mathbf{A} \cdot \mathbf{B}) = \mathbf{A} \cdot \frac{d\mathbf{B}}{du} + \frac{d\mathbf{A}}{du} \cdot \mathbf{B}$$

$$\frac{d}{du}(\mathbf{A} \wedge \mathbf{B}) = \mathbf{A} \wedge \frac{d\mathbf{B}}{du} + \frac{d\mathbf{A}}{du} \wedge \mathbf{B}$$

The remainder of this chapter is devoted to examples of the use of vectors to describe familiar physical quantities.

Work

The work done by a force acting at a point on a body is defined as the product of the force and the distance moved by the point of application in the direction of the force. Work has no directional property and so is a scalar.

Referring to Fig. 2.13, the work done through the displacement **s** by the constant force **F** is

$$W = F \times s \cos \theta$$

which is equal to $\mathbf{F} \cdot \mathbf{s}$, the scalar product of **F** and **s**.

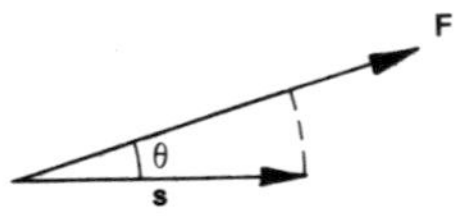

FIG. 2.13

Forces and Moments

Various elementary approaches are made to the definition of moment (of force or any other localised vector), so it is probably best in this context to go straight to the general definition.

The moment **M** of a force **F** about any point A is

$$\mathbf{M} = \mathbf{r} \wedge \mathbf{F} \tag{2.6}$$

where **r** is the vector from the point A to *any* point on the line of action of the force. The vector **M** is perpendicular to the plane containing the force and the point A, and of course obeys the right-hand screw rule. It can be seen in Fig. 2.14 that the perpendicular distance from the point A to the force **F** is r sin θ, wherever **r** is taken along the line of **F**.

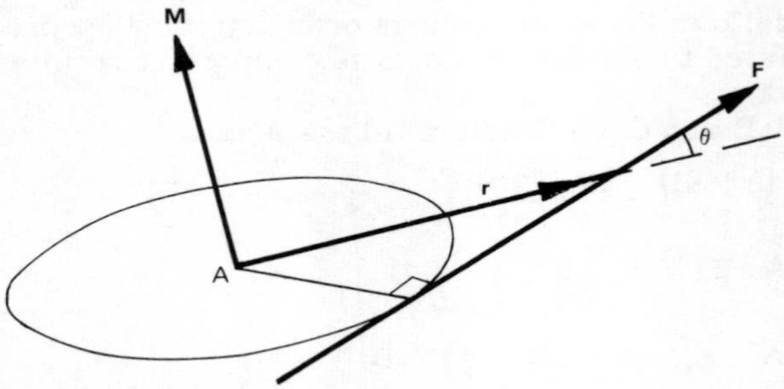

FIG. 2.14 Moment of force

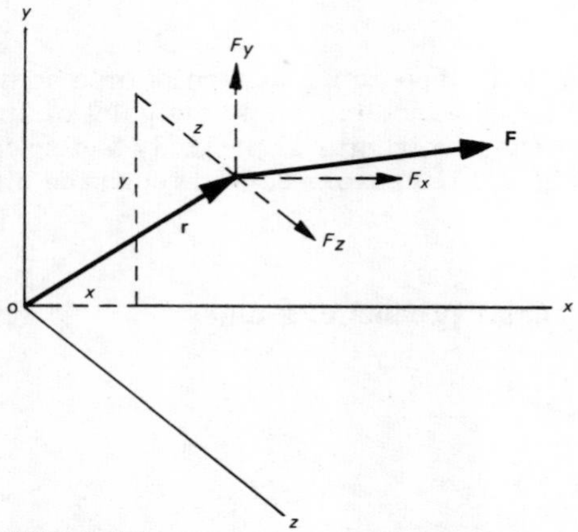

FIG. 2.15

If we now describe **F** in an *xyz* frame and take moments about the origin, we shall see a familiar standard result. In fig. 2.15, vector **r** ($= x\mathbf{i} + y\mathbf{j} + z\mathbf{k}$) gives the position of a point on the line of **F**($= F_x\mathbf{i} + F_y\mathbf{j} + F_z\mathbf{k}$). Then the moment of **F** about O is

$$\mathbf{M} = \mathbf{r} \wedge \mathbf{F}$$

which by our earlier results (2.3) and (2.4) can be written

$$\mathbf{M} = \mathbf{r} \wedge \mathbf{F} = \begin{vmatrix} \mathbf{i} & \mathbf{j} & \mathbf{k} \\ x & y & z \\ F_x & F_y & F_z \end{vmatrix} \tag{2.7}$$

$$= \mathbf{i}(yF_z - zF_y) + \mathbf{j}(zF_x - xF_z) + \mathbf{k}(xF_y - yF_x)$$

The last form of the moment may be familiar to the reader as the sum of the moments about the x, y, z axes of the components of the force **F**. You can easily confirm that by reference to Fig. 2.15.

Another way of expressing that result in words is to say that the three terms on the right-hand side are the cartesian components of the moment **M**, because **M** is a vector that can be resolved into components like any other vector.

$$\mathbf{M} = M_x\mathbf{i} + M_y\mathbf{j} + M_z\mathbf{k}$$

where $M_x = (yF_z - zF_y)$, $M_y =$ etc.

We shall see more of all this in Chapter 4, but one more term we should remind ourselves of is *couple*.

A couple can be made up of a pair of forces of equal magnitude acting along parallel lines, but in opposite directions.

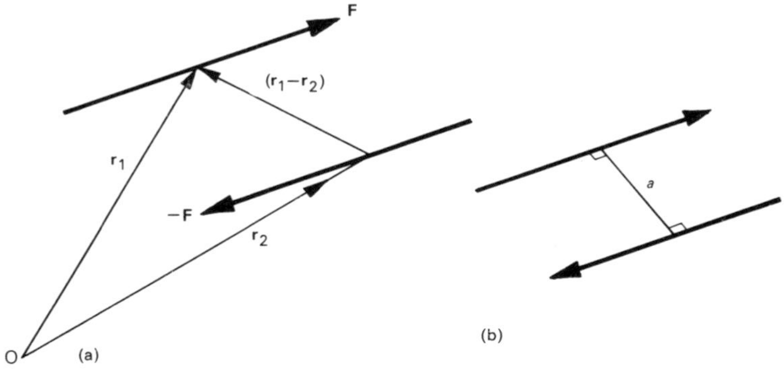

FIG. 2. 16 Couple

The total moment about O of the equal and opposite forces in Fig. 2.1 (a) is

$$\mathbf{r_1} \wedge \mathbf{F} - \mathbf{r_2} \wedge \mathbf{F}$$

which equals

$$(\mathbf{r_1} - \mathbf{r_2}) \wedge \mathbf{F}$$

This is independent of the location of O, depending only on the forces and the distance between them. Its magnitude is simply the product of the magnitude of the parallel forces and the perpendicular distance between them, aF, as in Fig. 2.16 (b).

A very useful result involving a couple is the 'shift of a force parallel to itself'. Figure 2.17 shows how adding two equal and opposite forces in the same line (i.e. adding nothing in total) leads to the result that a force **P** can be shifted parallel to itself through a distance b by the addition of a couple of magnitude bP, with axis perpendicular to the plane in which the force is shifted.

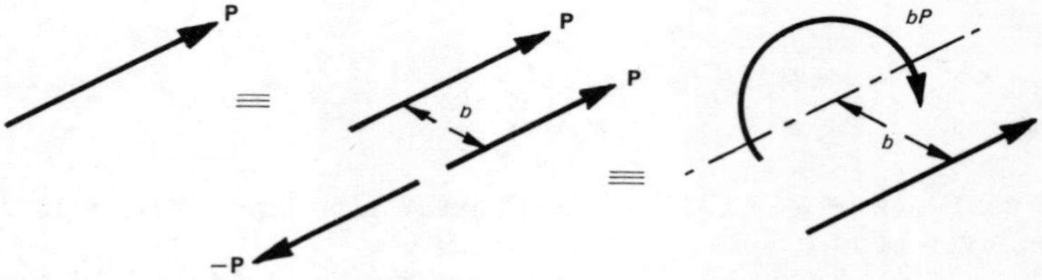

FIG. 2.17 Shifting a force parallel to itself

Velocity and Acceleration

Finally we take a brief look at velocity and acceleration; they are features of the subject called 'kinematics', which will be examined much more fully in Chapter 5, but as our first examples of differentiation of vectors, they are worthy of an introduction here.

The *position* of a point P can be defined by a vector **r** from any prescribed origin. As P moves in space the vector **r** changes in length and/or direction, as in Fig. 2.18. The *velocity* of P is defined as the time rate of change of position, denoted by d**r**/dt, the differential or derivative of **r** with respect to time. Referring to Fig. 2.18 (b), in which the point is shown at P at time t, and at P′ at time $t + \delta t$, the velocity v is

$$\mathbf{v} = \underset{\delta t \to 0}{\mathrm{Lt}} \frac{\delta \mathbf{r}}{\delta t} = \frac{\mathrm{d}\mathbf{r}}{\mathrm{d}t}$$

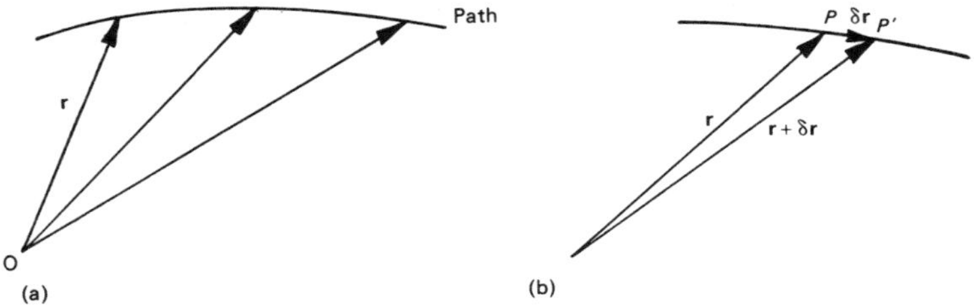

FIG. 2.18

The direction of **v** is tangent to the *path*, which is the locus of the position vector **r**.

The *acceleration* of P is defined as the time rate of change of velocity. It can be visulaised by sketching a diagram showing the variation of the velocity vector, as in Fig. 2.19.

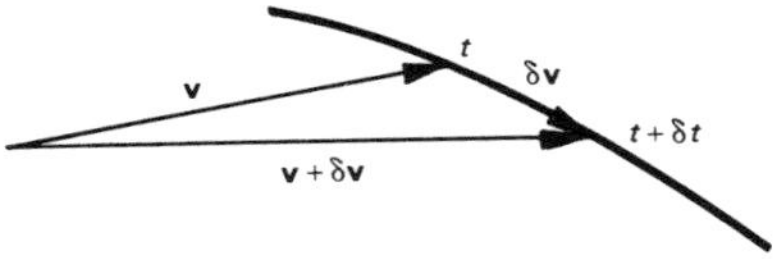

FIG. 2.19

$$\text{Acceleration } \mathbf{a} = \underset{\delta t \to 0}{\text{Lt}} \frac{\delta \mathbf{v}}{\delta t} = \frac{\mathrm{d}\mathbf{v}}{\mathrm{d}t} = \frac{\mathrm{d}^2\mathbf{r}}{\mathrm{d}t^2}$$

The acceleration vector is clearly tangent to the locus of the velocity vector, *not* to the path of the point P. It must be emphasised that P will have an acceleration by virtue of a change in *magnitude or direction* of **v**, so the acceleration cannot be zero if the point is moving along a curved path. This is illustrated by describing the acceleration in terms of its *normal* and *tangential* components. If $\hat{\mathbf{n}}$ and $\hat{\mathbf{t}}$ are normal and tangential unit vectors (with constant magnitude of unity, but varying in direction along the path), then we can see in Fig. 2.20 that the velocity **v** can be written

$$\mathbf{v} = v\hat{\mathbf{t}}$$

where v is the magnitude of the velocity (commonly called 'speed').

Now $$\mathbf{a} = \frac{\mathrm{d}\mathbf{v}}{\mathrm{d}t} = \dot{\mathbf{v}}$$

using the dot notation for differentiation with respect to time, so

$$\mathbf{v} = \frac{\mathrm{d}}{\mathrm{d}t}(v\hat{\mathbf{t}}) = \dot{v}\hat{\mathbf{t}} + v\dot{\hat{\mathbf{t}}} \qquad (2.8)$$

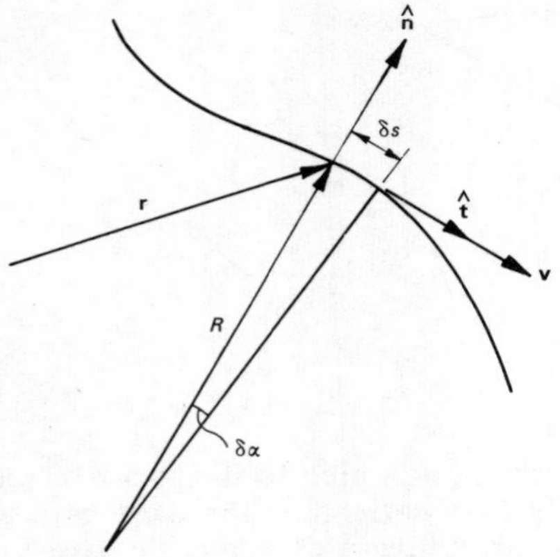

FIG. 2. 20

What does $\dot{\hat{\mathbf{t}}}$ mean? It is the time rate of change of the unit vector $\hat{\mathbf{t}}$, and can arise only from changing direction, since its magnitude is constant. It is clear from Fig. 2. 21 that changes in $\hat{\mathbf{n}}$ and $\hat{\mathbf{t}}$ must always be perpendicular to the current directions of $\hat{\mathbf{n}}$ and $\hat{\mathbf{t}}$ themselves. In particular $\delta\hat{\mathbf{t}}$ is in the direction $-\hat{\mathbf{n}}$. Its magnitude can be expressed as the small angle $\delta\alpha$ times unity, the length of $\hat{\mathbf{t}}$.

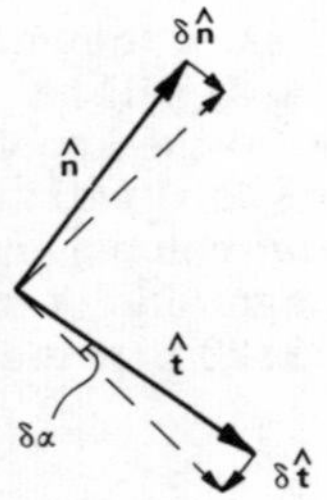

FIG. 2. 21

$$\delta\hat{\mathbf{t}} = \delta\alpha(-\hat{\mathbf{n}})$$

Similarly

$$\delta\hat{\mathbf{n}} = \delta\alpha(\hat{\mathbf{t}})$$

The angle $\delta\alpha$ is shown also in Fig. 2.20 and is the angle subtended at the instantaneous centre of curvature of the path, by the small arc δs, traced out between times t and $t + \delta t$.

Now $\qquad \delta s = v\delta t$

and so $\qquad \delta\alpha = \dfrac{v\delta t}{R}$, R being the instantaneous radius of curvature of the path.

Thus $\qquad \delta\hat{\mathbf{t}} = -v\dfrac{\delta t}{R}\hat{\mathbf{n}}$

and $\qquad \dot{\hat{\mathbf{t}}} = \mathrm{L_t}\dfrac{\delta\hat{\mathbf{t}}}{\delta t} = -\dfrac{v}{R}\hat{\mathbf{n}}$

Substituting in (2.8) we obtain at last

$$\dot{\mathbf{v}} = \dot{v}\hat{\mathbf{t}} - \frac{v^2}{R}\hat{\mathbf{n}} \tag{2.9}$$

Therefore the acceleration vector can be resolved into the two components: $\dot{v}$ tangential to the path, and $-v^2/R$ normal to the path; the negative sign indicates that the normal component is always directed inward towards the centre of curvature. The result (2.9) applies to plane or space curves. Resolution of acceleration into other components will be looked at in Chapter 5.

The principal aims in this chapter have been to become familiar with the idea of representing quantities such as force, position and velocity by vectors; to introduce the products of vectors, a scalar product of two vectors such as work and a vector product of two vectors such as the moment of a force.

This short introduction to vectors will be ample for the remainder of this book. Several applications will arise in later chapters, but the important thing at this stage is to try to think in vectors. Imagine deriving (2.9) in three dimensions without using vectors—you have probably done it before in two dimensions. The examples below will help to bring some of the points home.

Bibliography

1. Rutherford, D. E., *Vector Methods*, Oliver and Boyd, **1957**
2. Simons, S., *Vector Analysis*, Pergamon Press, **1964**

Examples

Ex. 2.1 By inspection, what are the angles between the following pairs of directions:

(a) $\mathbf{i}$ and $\mathbf{j}$
(b) $3\mathbf{i} + 4\mathbf{j}$ and $\mathbf{k}$
(c) $\mathbf{i} + \mathbf{j}$ and $\mathbf{i}$

Ex. 2.2 What is the angle between $\mathbf{i} + \mathbf{j} + \mathbf{k}$ and its projection in the xy plane? Remember $\mathbf{A} \cdot \mathbf{B} = AB\cos\theta$. [35°15']

Ex. 2.3 For the vectors $\mathbf{A} = \mathbf{i} + 2\mathbf{j} + 3\mathbf{k}$ and $\mathbf{B} = 4\mathbf{i} + 5\mathbf{j}$, find

(a) $\mathbf{A} + \mathbf{B}$

(b) $\mathbf{A} - \mathbf{B}$

(c) $\mathbf{A} . \mathbf{B}$ [14]

(d) $\mathbf{A} \wedge \mathbf{B}$ $[-15\mathbf{i} + 12\mathbf{j} - 3\mathbf{k}]$

(e) the angle between **A** and **B** [54° 15']

Ex. 2.4 Find the magnitude and direction of the moment about the origin, of a force of 1 000 N (newtons) which passes through the point $x = 1\text{m}, y = 2\text{m}, z = 3\text{m}$ in the direction $3\mathbf{i} + 2\mathbf{j} + \mathbf{k}$. [2619 N m, $-\mathbf{i} + 2\mathbf{j} - \mathbf{k}$]

Ex. 2.5 If the sum of two vectors has the same magnitude as their difference, show that the two vectors are perpendicular.

3

Newton's Laws of Motion

'The discovery and formulation of the laws of mechanics, like most great discoveries, was not the work of one man alone, but Newton's contribution is so preponderant that the laws are called by his name.'[1]

J. P. Den Hartog

In the first half of this chapter Newton's laws of motion and gravitation will be presented briefly and with little comment. Between them they constitute the foundation of all non-relativistic mechanics, and certainly the whole of engineering mechanics comes very safely into that category. Nearly everyone who reads this book will have encountered the modern statements of Newton's laws of motion, so out of historical interest (and Newton's contribution to the subject is by far the outstanding historical landmark) the laws are quoted below in the form of an early translation of Newton's own words in Latin, along with his own notes of explanation of the laws.

In the second half of the chapter we shall meet the internationally agreed system of units of measurement and dimensions, the SI system, and the simple but valuable technique of 'dimensional analysis'.

Much has been written about the origins and the consequences of the formulation of Newton's laws, by supporters and by sceptics; about the early work of Galileo and Huygens, before Newton, and the many giants of the eighteenth century after Newton. This little book is not the place to give an account of these writings, but for those interested there is much fascinating reading in the references given at the end of this chapter.

There is no better way to present the laws first than to quote Newton himself. The next page or so, giving the laws and explanatory illustrations, is a direct quotation from the *Principia*, published in 1687, as translated by Motte in 1729 and revised by Cajori for publication in 1934 [2]. It should be noted that what is now called 'momentum' is called 'motion' in the translation below, and defined by Newton 'The quantity of motion is the measure of the same, arising from the velocity and quantity of matter conjointly.'

Here begins the quotation from the *Principia*.

Axioms, or laws of motion

Law I

Every body continues in its state of rest, or of uniform motion in a right line, unless it is compelled to change that state by forces impressed upon it.

Projectiles continue in their motions, so far as they are not retarded by the resistance of the air, or impelled downwards by the force of gravity. A top, whose parts by their cohesion are continually drawn aside from rectilinear motion, does not cease its rotation, otherwise than it is retarded by the air. The greater bodies of the planets and comets, meeting with less resistance in freer spaces, preserve their motions both progressive and circular for a much longer time.

Law II

The change of motion is proportional to the motive force impressed; and is made in the direction of the right line in which that force is impressed.

If any force generates a motion, a double force will generate double the motion, a triple force triple the motion, whether that force be impressed altogether and at once, or gradually and successively. And this motion (being always directed the same way with the generating force), if the body moved before, is added to or subtracted from the former motion, according as they directly conspire with or are directly contrary to each other; or obliquely joined, when they are oblique, so as to produce a new motion compounded from the determination of both.

Law III

To every action there is always an equal reaction: or, the mutual actions of two bodies upon each other are always equal, and directed to contrary parts.

Whatever draws or presses another is as much drawn or pressed by that other. If you press a stone with your finger, the finger is also pressed by the stone. If a horse draws a stone tied to a rope, the horse (if I may so say) will be equally drawn back towards the stone; for the distended rope, by the same endeavour to relax or unbend itself, will draw the horse as much towards the stone as it does the stone towards the horse, and will obstruct the progress of the one as much as it advances that of the other. If a body impinge upon another, and by its force change the motion of the other, that body also (because of the equality of the mutual pressure) will undergo an equal change, in its own motion, towards the contrary part. The changes made by these actions are equal, not in the velocities but in the motions of the bodies; that is to say, if the bodies are not hindered by any other impediments. For, because the motions are equally changed, the changes of the velocities made towards contrary parts are inversely proportional to the bodies. This law takes place also in attractions, as will be proved in the next Scholium.

There ends the quotation.

The subject of dynamics was developed throughout the eighteenth century, and the laws were not subjected to adverse criticism until late in the nineteenth century. Then, notably by Mach [3] and Hertz [4], they were critically examined, largely on their form of presentation, rather than their fundamental correctness. That early anxiety was replaced by scepticism in the first flush of relativity theory; for instance, Eddington reformulated the First Law, 'Every body continues in its state of rest or uniform motion in a straight line except in so far as it doesn't' [5].

Eventually, Newtonian dynamics was put into perspective by Sir James Jeans [6], who, describing the *Principia* as 'certainly the greatest scientific work ever produced by the human intellect', recalled that Mach in 1883 had pointed out that Newton's definitions were not really definitions but assumptions which specified the particular kind of system to which the theorems of the *Principia* apply. That is how we should regard them now: except at the very extremes of scale and velocity, Newton's laws *work* in that they give extremely accurate descriptions of dynamical phenomena; they can be used without reservation in engineering mechanics.

Most modern statements of the laws leave the first and third unaltered except for minor changes in wording; the second takes one of the two forms:

Second Law:
'The time rate of change of momentum of a body is proportional to the force producing it, and the change takes place in the direction in which the force is acting';

OR
'A body on which a force is acting experiences an acceleration in the direction of that force, proportional to the force and inversely proportional to the mass of the body.'

These are plainly equivalent for a body of constant mass. We shall be taking a look at bodies of changing mass later. If a system of units is adopted in which the factor of proportionality is unity, these statements can be written

$$\mathbf{F} = \frac{\mathrm{d}}{\mathrm{d}t}(m\mathbf{v}) \text{ and } \mathbf{F} = m\mathbf{a} \tag{3.1}$$

So far we have had no discussion about the nature of force and mass, a discussion which occupies a good deal of the writing in the references at the end of the chapter. It is not helpful to assert that force is that which changes motion.

The natural approach for ordinary mortals is through the human experiences of force, particularly weight and muscular force. We all know about muscular force because we are what we are, and about weight because we live on the earth. A straight forward concept of direct forces like muscular forces is of an action which deforms a spring, provided of course there is an appropriate constraint to keep the spring at rest, in other words a reaction.

Weight on the other hand, although a matter of common experience, and measurable with great accuracy in the laboratory, is not so easy to comprehend. It is, however, easily *described* (as the force pulling a body towards the earth), thanks once again to Newton. It was he who first appreciated that the same kind of force which keeps a planet in orbit round the sun causes bodies to fall to the ground on earth. Newton's Law of Gravitation is

$$F = G\frac{m_1 m_2}{r^2} \tag{3.2}$$

where F = the mutual gravitational attractive force between two particles,
m_1, m_2 = the masses of the two particles,
G = a universal constant called the gravitational constant,
r = the distance between the particles.

'Particle' is used here in its usual sense, denoting a body with finite mass but negligible dimensions; in this case negligible compared with r.

For a spherical body with radial symmetry (a uniform hollow sphere, for instance, or a solid sphere whose density varies only with the radius), it can be shown that the gravitational attraction is the same as for a particle of the same mass at the centre of the sphere. The *weight* of a body is the gravitational force exerted on the body by the earth; if we imagine for the moment that the earth is a sphere of the kind described in the last sentence, we would expect the weight of a body to be the same all over the surface of the earth, but to decrease if it is moved outwards from the surface of the earth, to a quarter of its surface weight for example at an altitude equal to the radius of the earth (doubling r in (3.2)).

The expressions (3.1) and (3.2) can be combined to give the acceleration due to gravity of a body falling towards the earth (*in vacuo,* strictly),

$$g = \frac{GM}{r^2} \tag{3.3}$$

where M is the mass of the earth.

Actually the weight of a body and the local value of g vary very slightly from place to place on the earth's surface. There are two reasons: one is that the earth is neither perfectly spherical nor quite radially symmetrical, the other is that the earth is rotating (the latter effect will, I hope, be understood after a reading of Chapter 5). The variations are small and can be neglected in engineering calculations.

We are left then with the mass of a body as the property which resists or limits the acceleration caused by a force, and as the property which determines the body's attraction for, and to, other bodies. In both these actions the mass alone is significant; the chemical constitution is irrelevant. It is mass which is the intrinsic property of the body. Having said that, mass is assigned to a body by comparison with an arbitrarily selected standard of mass; the comparison is usually made by measuring the weight, either on a calibrated spring or by direct comparison with reference masses on a balance. Although the weights of all bodies depend on both position and reference frame, their ratios do not but are equal to the ratios of their masses, so comparison of weights provides an accurate determination of mass.

> 'Until a new maestro emerges, or perhaps until space travel provides new observational data on our cosmic environment, the blueprint of the universe remains essentially the one that Newton drew for us, in spite of all disturbing rumours about the curvature of space, the relativity of time, and the runaway nebulae.'
>
> Arthur Koestler

Units of measurement and dimensions

Many systems of units have been used by scientists and engineers in the past, but mercifully the miscellany of systems, and the confusion caused by some of them which were incoherent, are behind us, as the world (the technical world at any rate) moves slowly but firmly towards the Système International d'Unités, SI for short. The basic SI units are

Quantity	Unit	Symbol
length	metre	m
mass	kilogramme	kg
time	second	s
electric current	ampere	A
thermodynamic temperature	degree Kelvin	°K
luminous intensity	candela	cd

Any other physical quantity can be measured by a unit derived from this basic set. In mechanics all quantities can be expressed in terms of the basic units of length, mass and time and the units derived from them, supplemented by the radian (which is dimensionless) as the unit of plane angle.

There are, of course, many derived units, some of them having special names. The most important in mechanics is the unit of force; it is given by Newton's second law as the force which imparts unit acceleration (1 m/s^2) to unit mass (1 kg), and is appropriately named the 'newton'. The unit of energy called the 'joule' is the work done when the point of application of a force of 1 newton is displaced through a distance of 1 metre in the direction of the force. The unit of power called the 'watt' is equal to one joule per second.

Some of the derived SI units are:

Quantity	Unit	Symbol
force	newton	N = kg m/s^2
work, energy	joule	J = N m
power	watt	W = J/s
frequency	hertz	Hz = 1/s
angular velocity		rad/s
velocity		m/s
acceleration		m/s^2
density		kg/m^3
pressure and stress		N/m^2
momentum		kg m/s
moment of inertia		kg m^2

It is useful to distinguish between dimensions and units. In the table of basic SI units above, the quantities in the left-hand column are dimensions, and those in the second column are units. Useful results can often be derived from a consideration of dimensions alone, without reference to any system of units; indeed depending on the fact that a physical relation must be independent of the units used, being correct in any consistent system, metre-kilogramme-second or furlong-ton-fortnight.

All quantities in mechanics can be expressed in terms of the three fundamental quantities, mass, length and time, denoted as dimensions (not units) by M, L and T, plus an angle to describe rotation, which is dimensionless. A physical relation must be dimensionally homogeneous, so a useful check on any derived relation is that it is at least *dimensionally* correct. Of course that is not a complete check of its correctness but it is a necessary condition. For example we shall see later that the kinetic energy of a body of mass m moved from rest to a velocity v by the action of a constant force F applied through a distance s is equal to the work done by the force:

$$\tfrac{1}{2}\, mv^2 = Fs$$

Remembering that force has the dimensions mass × acceleration, we can write the dimensions of the expression

$$M \times \left(\frac{L}{T}\right)^2 = M \times \left(\frac{L}{T^2}\right) \times L$$

$$\frac{ML^2}{T^2} = \frac{ML^2}{T^2}$$

which confirms that the expression is dimensionally homogeneous. Note that the ½, which is a true constant, was left out of the dimensional equation as any constant coefficient would be. That would *not* apply to an index such as the 'squared' in v^2, which raises the power of one or more dimensions. The index, or any exponent, or the argument in any transcendental term like a logarithm or trigonomeric term, must itself be dimensionless.

A more complicated example is given by the flow of fluid through a pipe. The volume rate of laminar flow of a fluid of viscosity η through a pipe length of l and radius r is

$$V = \frac{\pi r^4}{8\eta}\left(\frac{p_1 - p_2}{l}\right) \tag{3.4}$$

where p_1 and p_2 are the pressures at the upstream and downstream ends of the length l. V is a volume per unit time so its dimensions are L^3/T. r and l are lengths and obviously have dimension L. The p's are pressures and are therefore force per unit area; force has dimensions ML/T^2 so pressure is M/LT^2. Finally the viscosity is given by the law first propounded by Newton (who else ?)

$$\text{shear stress in fluid} = \text{viscosity} \times \text{velocity gradient}$$

$$\text{viscosity} = \frac{\text{stress}}{\text{velocity} \div \text{distance}}$$

$$= \frac{M}{LT^2} \div \frac{L}{TL} = \frac{M}{LT}$$

In the pipe flow expression therefore

$$\frac{L^3}{T} = L^4 \times \frac{LT}{M} \times \frac{M}{LT^2} \times \frac{1}{L} = \frac{L^3}{T}$$

which checks correctly.

Incidentally, a similar check on the *units* in a numerical calculation is often worthwhile.

In addition to this sort of check of dimensional homogeneity, the idea of using dimensions as quantities has been exploited in the theory of *dimensional analysis* and *model theory*. Model theory is widely used in interpreting model experiments in wind-tunnels, model rivers and estuaries and the like, but is not really our subject here. Dimensional analysis is so useful in mechanics generally that the principle and the technique will be briefly outlined below. An account of the method can be found in many books on fluid mechanics and in some texts devoted entirely to the subject.

Dimensional Analysis

The purpose of a dimensional analysis is effectively to reduce the number of variables active in a physical situation by grouping them together into dimensionless parameters. This leads to economy in experimental or computational work, and facilitates correlation and presentation of results. It is normally used only in connection with problems for which a full theoretical solution cannot be found.

The first step is to decide what variables have an influence on the problem. It is necessary therefore to have some understanding of the physical mechanism of the problem, based perhaps on the governing equations (which can be formulated but not solved) or on experience of similar problems.

If the variables which influence a particular system are $v_1, v_2, \ldots, v_n$, the equation describing the behaviour of the system can in principle be written

$$f\ (v_1, v_2, \ldots, v_n) = 0$$

where f is some unknown function of the variables. It can be shown (Buckingham's so-called π-theorem) that this expression can be recast in the form

$$\phi\ (\pi_1, \pi_2, \ldots, \pi_{n-k}) = 0$$

where the π's are dimensionless parameters formed from the products of powers of the variables v_1, etc. Note that the number of such independent π terms will be n-k. k is the number of fundamental quantities (M, L and T) involved in all the v's. In other words the number of variables will be reduced by three if all of M, L and T are involved in the formulation of the problem.

The best way to get the hang of the technique is to look at a few examples. First note a few rules and the steps to be taken in one of the several approaches to the dimensionless groups.

1. Decide what independent variables are relevant—n in number.
2. Number of fundamental dimensions involved—k
3. The system can therefore be described by n-k dimensionless parameters.
4. Pick out k of the n variables as primary variables. These must all have different dimensions, and between them they must contain all the fundamental dimensions involved.
5. Express the remaining n-k variables as π_1, π_2, etc.
6. Each π term must include at least one variable which does not appear in any other of the π terms.
7. Any of the π terms can be multiplied by any of the other π terms, or any power thereof.

Consider first the trivial case of a mass m moving in a circle of radius r with constant velocity v. What can dimensional analysis tell us about the force P required to sustain the motion?

The variables are	P	m	r	v
Their dimensions are	$\frac{ML}{T^2}$	M	L	$\frac{L}{T}$

There are four variables and three dimensions. Therefore we expect one dimensionless group. Select m, r and v as primary variables.

Then

$$\pi_1 = \frac{P}{m^{\alpha} r^{\beta} v^{\gamma}}$$

For this to be dimensionless,

$$\frac{ML}{T^2} = M^{\alpha} L^{\beta} \left(\frac{L}{T}\right)^{\gamma}$$

Now equate powers of M, L and T:

$$\begin{aligned} M&: \ 1 = \alpha \\ L&: \ 1 = \beta + \gamma \\ T&: \ -2 = -\gamma \end{aligned}$$

From which, plainly, $\alpha = 1, \beta = -1, \gamma = 2$.
The group we seek is therefore

$$\pi_1 = \frac{Pr}{mv^2}$$

Dimensional analysis tells us therefore that the equation governing the motion is of the form

$$\phi\left(\frac{Pr}{mv^2}\right) = 0$$

and no more. In this case, however, intuition tells us (pretending that we do not already know the answer from a complete analysis) that there is only *one value* of P for given values of m, r and v, so the relation must be

$$\frac{Pr}{mv^2} = \text{a constant}$$

Now to do something more realistic, but taking once again a problem whose answer we know from analysis, let us see how far we can get towards the expression (3.4) for viscous flow through a pipe. The variables and their dimensions are

V	η	r	l	Δp
$\frac{L^3}{T}$	$\frac{M}{LT}$	L	L	$\frac{M}{LT^2}$

Δp is the pressure difference over the length l.

There are five variables and three dimensions, so we expect two dimensionless groups. Select η, r and Δp as primary variables. Why? you are saying to yourself. When we have finished the example we will discuss it.

$$\pi_1 = \frac{V}{\eta^{\alpha_1} r^{\beta_1} \Delta p^{\gamma_1}}$$

$$\frac{L^3}{T} = \left(\frac{M}{LT}\right)^{\alpha_1} L^{\beta_1} \left(\frac{M}{LT^2}\right)^{\gamma_1}$$

$$\left.\begin{array}{l} M\text{:}\ \ 0 = \alpha_1 + \gamma_1 \\ L\text{:}\ \ 3 = -\alpha_1 + \beta_1 - \gamma_1 \\ T\text{:}\ \ -1 = -\alpha_1 - 2\gamma_1 \end{array}\right\} \qquad \begin{array}{l} \alpha_1 = -1 \\ \beta_1 = 3 \\ \gamma_1 = 1 \end{array}$$

$$\text{So } \pi_1 = \frac{V\eta}{r^3 \Delta p}$$

$$\pi_2 = \frac{l}{\eta^{\alpha_2} r^{\beta_2} \Delta p^{\gamma_2}}$$

$$L = \left(\frac{M}{LT}\right)^{\alpha_2} L^{\beta_2} \left(\frac{M}{LT^2}\right)^{\gamma_2} \tag{3.16}$$

$$\left.\begin{array}{l} M\text{:}\ \ 0 = \alpha_2 + \gamma_2 \\ L\text{:}\ \ 1 = -\alpha_2 + \beta_2 - \gamma_2 \\ T\text{:}\ \ 0 = -\alpha_2 - 2\gamma_2 \end{array}\right\} \qquad \begin{array}{l} \alpha_2 = 0 \\ \beta_2 = 1 \\ \gamma_2 = 0 \end{array}$$

$$\text{So } \pi_2 = \frac{l}{r}$$

The dimensional analysis tells us therefore that the equation describing the flow-rate will be of the form

$$\phi\left(\frac{V\eta}{r^3 \Delta p}, \frac{l}{r}\right) = 0$$

or this could be written

$$\frac{V\eta}{r^3 \Delta p} = f\left(\frac{l}{r}\right)$$

or

$$V = \frac{r^3 \Delta p}{\eta} f\left(\frac{l}{r}\right)$$

To make further progress (still pretending that we do not know the answer, and cannot do the fluid-flow theory) we must now do an experiment. A suitable experiment would tell us that with r, Δp and η constant, the flow-rate was inversely proportional to the length of the pipe.

i.e. $$f\left(\frac{l}{r}\right) = \text{constant} \times \frac{r}{l}$$

So $$V = \text{constant} \times r^4 \frac{\Delta p}{\eta l}$$

The experiment would also tell us the constant if we could measure η, the viscosity of the fluid.

Now, finally, we return to the question above, why did we pick those particular primary variables? In general the answer contains a number of elements. If you have some experience of dimensional analysis and know a fair amount about the problem you are analysing, you may have some idea about the form you would like the groups to take. Also it is common sense not to pick the dependent variable (in this case V) as one of the primaries, because it might then occur in several of the π groups. But because of number 7 in the list of rules and steps, there is freedom to do any amount of manipulation of the π terms among themselves, provided of course the correct total number of groups remains. In addition the conditions of rule 4 must be observed. If the rules are broken, the analysis will show it by producing a nonsense result such as $1 = 0$.

This is the briefest possible outline, and there is much more to the whole story; including the use of other basic dimensions, particularly the use of force as a basic quantity in static problems, where time is not an active variable.

Bibliography

1. Den Hartog, J. P., *Mechanics*, McGraw-Hill, 1948, Dover, **1961**
2. Newton, I. *Mathematical Principles of Natural Philosophy (The Principia)* 1687, Cajori's revision of Motte's (1729) translation, University of California Press, **1960**
3. Mach, E., *The Science of Mechanics*, 1883, published in English by the Open Court Publishing Co., **1902**
4. Hertz, H., *The Principles of Mechanics*, first English translation 1899, now Dover, **1956**
5. Eddington, A. S., *The Nature of the Physical World*, Cambridge University Press, **1928**
6. Jeans, J. H., *The Growth of Physical Science*, Cambridge University Press, 1947
7. Hill, R., *Principles of Dynamics*, Pergamon, **1964**
8. Kenyon, R. A., *Principles of Fluid Mechanics*, Ronald Press, **1960**
9. Huntley, H. E., *Dimensional Analysis*, Macdonald, **1952**

Examples

3.1 In an apparatus to measure the viscosity of fluids, a vertical cylindrical tube is filled with fluid and a steel ball, whose radius a is only slightly smaller than that of the tube, $a + c$, falls slowly through the fluid. By dimensional analysis, work out a set of dimensionless groupings of the velocity of descent V, the radius of the ball a, the radial clearance c, the viscosity η, the densities of the ball ρ and fluid ρ_0 and the acceleration due to gravity g.

3.2 A field gun fires a shell of mass m at muzzle velocity V and elevation θ. If the air-resistance is kv^2, where k is a constant and v the velocity in flight, and the acceleration due to gravity is g, express the range R as a function of dimensionless groupings of the other variables. Do it first neglecting air resistance and note that θ, being an angle, must appear in its own right as a dimensionless 'group'.

In the analysis including air resistance, the first step will be to find the dimensions of k.

$$\left[\frac{V}{\sqrt{ag}}, \frac{c}{a}, \frac{\zeta}{\zeta_0}, \frac{\eta}{\zeta\sqrt{ag^3}}\right] \qquad 3.1$$

$$\left[R = \frac{v^2}{g}\, \phi\,(\theta), R = \frac{v^2}{g}\, \phi\, \left(\theta, \frac{kv^2}{mg}\right)\right] \qquad 3.2$$

4 Statics

A body, or system of bodies, which is not being accelerated, is not acted upon by a resultant force. If any forces are acting upon it, they must all be in balance. This principle will be stated formally in the 'conditions of equilibrium', which will be applied to a number of real situations, employing and illustrating the vector methods described in Chapter 2. The applications will include simple structures and slowly moving mechanisms.

We shall meet, for the first time, methods based on work and energy, which greatly simplify the analysis of many systems.

Mechanics is the study of forces and motions, including displacements and deformations. Statics is concerned primarily with stationary bodies or assemblies of bodies, but it can also tackle moving assemblies so long as the *accelerations* are small; or, more precisely, so long as the (mass x acceleration)s are negligible compared with the forces applied to the assembly. The use of simple statics is not ruled out by changes of geometry, even by continuously changing geometry. In the absence of accelerations, the forces are in a state of balance or equilibrium at any instant, although they may be varying with time.

This chapter is largely an introduction to subjects which are not dealt with further in this book, but in specialist books covering the field of 'engineering statics', particularly stress analysis and structural analysis. But what we shall do in this chapter will be useful in itself and will be particularly helpful when we tackle the dynamics of bodies in the last chapter.

Unlike dynamics, statics is an ancient science going back to Archimedes; the modern formulation of the subject started with the work of Stevinus (1548-1620), a Dutch engineer. The thoughts of Stevinus about weights on inclined planes led to the development of what are now known as the 'parallelogram of forces' (first explicitly formulated by Newton and Varignon), the 'polygon of forces' and the 'funicular polygon', composition of forces and resolution of forces.

Two of the foundation stones of statics are the parallelogram of forces and Newton's third law (it would be worth turning back and reading Newton's third law and his notes about it again). A third is the *principle of transmissibility*, which can be stated: *a force may be considered to act at any point along its line of action as far as the conditions of equilibrium are concerned.* This principle must be employed with caution where non-rigid (i.e. real) bodies are involved, as Fig. 4.1 forcefully shows; but more of that later.

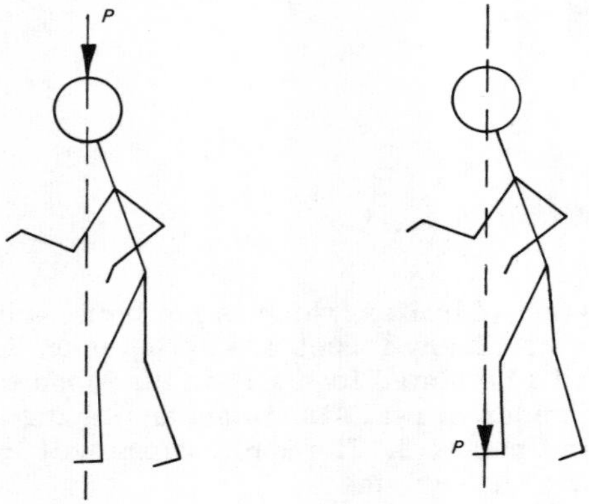

FIG. 4.1 Concerning the principle of transmissibility

Conditions of Equilibrium

Newton's laws refer implicitly to particles only, so we start from the point that a particle on which the resultant force is zero has no acceleration. The particle is said to be in *equilibrium* Since a particle is a mass-point, all the forces acting on it must necessarily intersect, so the single condition of equilibrium is that the resultant of all the forces acting on it must be zero.

To extend the equilibrium condition to a finite body, we must indulge in an act of faith—that of conceptually dividing the finite body into a number of particles, and applying the condition of equilibrium to each of the 'separate' particles. To quote Rodney Hill [1],

> 'Even when the considered continuum is only part of a naturally coherent body, the mechanical influence of the remainder is still conceived to be fully represented by a distributed traction over the interface. The local tractions of either part on the other are, of course, taken to be equal and opposite. This concept replaces, on the macroscopic level, what would be described in fine detail as field forces between atoms and electrons. In ultimate justification stands the whole of engineering science.'

Figure 4.2 shows a body which is in equilibrium under the action of several external forces. We imagine it to be made up of a large number of particles. The diagram shows only a few of the 'particles', every one of which is in equilibrium

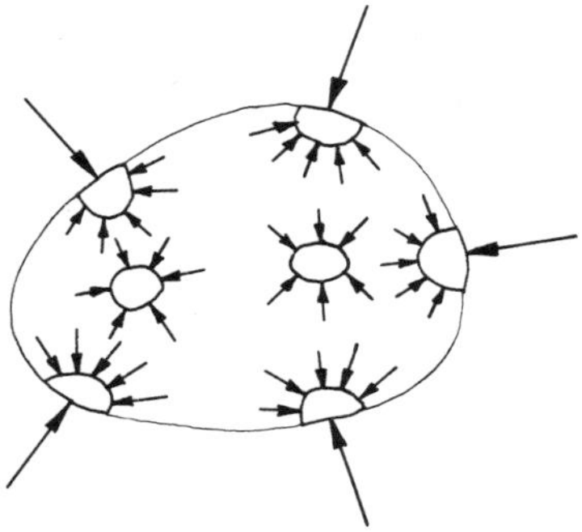

FIG. 4.2

under the action of a large number of forces, mostly internal, being pushes or pulls from neighbouring particles. A few of the particles are directly exposed to *external* forces, and they have the job of transmitting the forces into the body. The study of how the forces are transmitted into and through the body is the important subject called *stress analysis*.

Now it is clear that, however you imagine the body to be cut up into particles, the internal forces will all act in equal and opposite pairs (Newton 3) and will, in the overall condition of equilibrium for the entire body, cancel each other out. So the condition of equilibrium for the finite body is that all the *external* forces are in equilibrium. What the condition is we shall see in a moment, but first we must examine one rash assumption made in the last paragraph.

We have assumed that the body is *capable* of transmitting the forces without breaking or being distorted from its original shape. The classical subject of statics deals only with the transmission of forces by bodies whose changes of size and shape are negligible under the influence of the forces; hence its full title *rigid body statics*. Just what this assumption implies, and the limitations it imposes, we shall see presently, but now back to the condition of equilibrium.

The condition of equilibrium for the body in Fig. 4.2 is that the external forces should themselves be in equilibrium; i.e. their *resultant* should be zero. It can be shown (see Rutherford [2] for instance) that any system of forces in three dimensions can be compounded into a single resultant force and a single couple acting about a line parallel to the resultant force; this ultimate composition is called a *wrench*. For equilibrium both the force and the couple must be zero.

In a two-dimensional system of forces, all the forces act in one plane, so clearly the wrench reduces to *either* a force *or* a couple.

Our purposes, however, will be better served by compounding the force system into a couple and a single force at an *arbitrary point*, which will be the origin of coordinates. The single force will be the same as in the wrench, but the couple in general will not. Consider a set of forces $\mathbf{F}_1$ at $\mathbf{r}_1$, $\mathbf{F}_2$ at $\mathbf{r}_2$, etc. We have seen in Chapter 2 how a force can be shifted parallel to itself by the introduction of a couple. Suppose we introduce forces $\mathbf{F}_1$ and $-\mathbf{F}_1$ at the origin of coordinates O (see Fig. 4. 3). The $\mathbf{F}_1$ at $\mathbf{r}_1$ and the $-\mathbf{F}_1$ at O constitute a couple $\mathbf{r}_1 \wedge \mathbf{F}_1$, and so the force $\mathbf{F}_1$ at $\mathbf{r}_1$, is equivalent to a force $\mathbf{F}_1$ at O plus a couple $\mathbf{r}_1 \wedge \mathbf{F}_1$. So the whole system of forces is equivalent to the sum of $\mathbf{F}_1$, $\mathbf{F}_2$, etc., acting at O together with couples $\mathbf{r}_1 \wedge \mathbf{F}_1$, $\mathbf{r}_2 \wedge \mathbf{F}_2$, etc. They can be compounded into a single force $\Sigma\mathbf{F}$ and a single couple $\Sigma\mathbf{r} \wedge \mathbf{F}$. For equilibrium they must both be zero.

i.e. $$\Sigma\mathbf{F} = 0 \text{ and } \Sigma\mathbf{r} \wedge \mathbf{F} = 0 \tag{4.1}$$

In words, the conditions of equilibrium of a body are that the vector sum of the external forces must be zero and that the total moment of the external forces about an arbitrary point should be zero.

It should be noted that taking moments about more than one point merely provides an alternative presentation of the information in (4. 1); no additional information is provided.

The vector equations (4.1) are equivalent in a three-dimensional case to six scalar components, expressed in any convenient set of coordinates. For example the single force $\mathbf{F}$ applied at $\mathbf{r}$ can be expressed in cartesian components:

$$\mathbf{F} = F_x\mathbf{i} + F_y\mathbf{j} + F_z\mathbf{k}$$
$$\mathbf{r} = x\mathbf{i} + y\mathbf{j} + z\mathbf{k}$$

Its moment about the origin is

$$\mathbf{M} = \mathbf{r} \wedge \mathbf{F} = \begin{vmatrix} \mathrm{i} & \mathrm{j} & \mathrm{k} \\ x & y & z \\ F_x & F_y & F_z \end{vmatrix}$$

$$= (yF_z - zF_y)\mathbf{i} + (zF_x - xF_z)\mathbf{j} + (xF_y - yF_x)\mathbf{k}$$

You might like to check the last expression by taking moments about the three axes of the three components of $\mathbf{F}$, remembering to obey the right-hand screw rule.

Summed over all the external forces, these components are equated to zero to give the six conditions of equilibrium:

$$\left.\begin{aligned} \Sigma F_x = \Sigma F_y = \Sigma F_z = 0 \\ \Sigma M_x = \Sigma M_y = \Sigma M_z = 0 \end{aligned}\right\} \tag{4.2}$$

As a trivial example, examine the forces on an old-fashioned motor car starting-handle. The situation is described in Fig. 4. 4, and could be solved

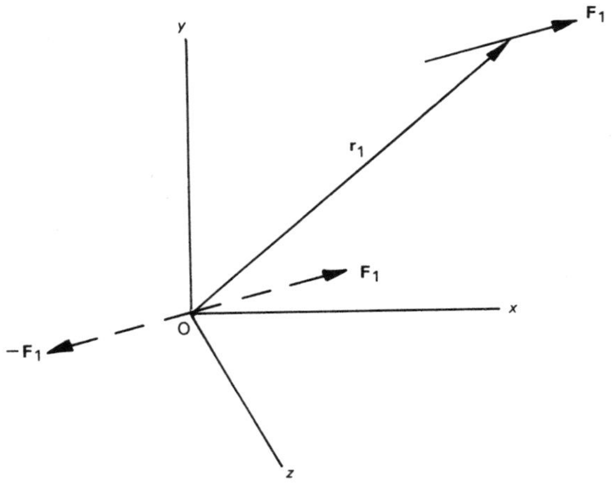

FIG. 4. 3

by inspection, but we will engage the full power of equations (4. 2).

T is the moment or torque required to 'turn the engine over', and F is the corresponding cranking force, applied manually. To keep the shaft of the starting-handle on the axis of the engine, a force $\mathbf{R}$ (components R_y and R_z) must be provided by the locating socket at the front of the engine, and a force P (P_y and P_z) is provided by a rudimentary bearing at the front of the car (probably just a hole in the bumper). All the forces shown will rotate with the starting-handle, so we can either let the y and z axes rotate also, or simply consider the thing when the handle lies in a horizontal plane and F is vertical. Applying (4. 2)

$$\begin{array}{ll} F_x & 0 = 0 \\ F_y & 0 = P_y + R_y \\ F_z & 0 = F + P_z + R_z \\ M_x & 0 = T - Fa \\ M_y & 0 = -P_z b - Fc \\ M_z & 0 = P_y b \end{array}$$

Solving,

$$P_y = R_y = 0; F = \frac{T}{a}; P_z = -\frac{Tc}{ab}; R_z = \frac{T}{a}\left(\frac{c}{b} - 1\right)$$

Note that these are the forces on the starting-handle, and are equal and opposite to the forces on the bearing, etc. To avoid confusion in this matter it is most helpful to *isolate* the body under consideration from supporting or connected bodies, and show only forces acting on the single body.

Figure 4.4 (b) shows the starting-handle *isolated* from the engine and the bearing. It is imperative that all the relevant internal as well as external forces are shown on the isolated body. A sketch such as Fig. 4.4 (b) is often called a 'free-body diagram'.

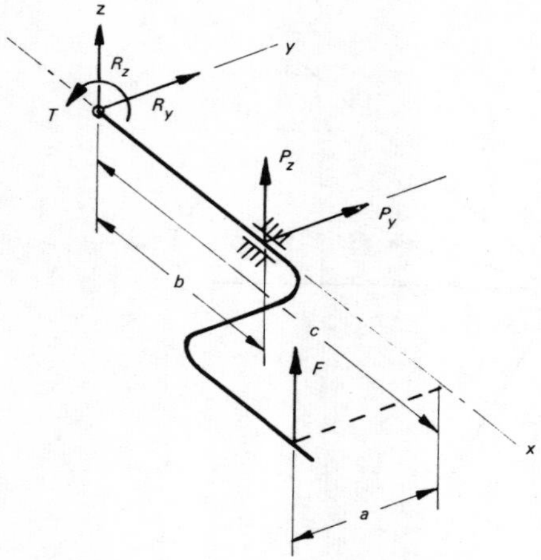

FIG. 4.4 (a) Starting handle

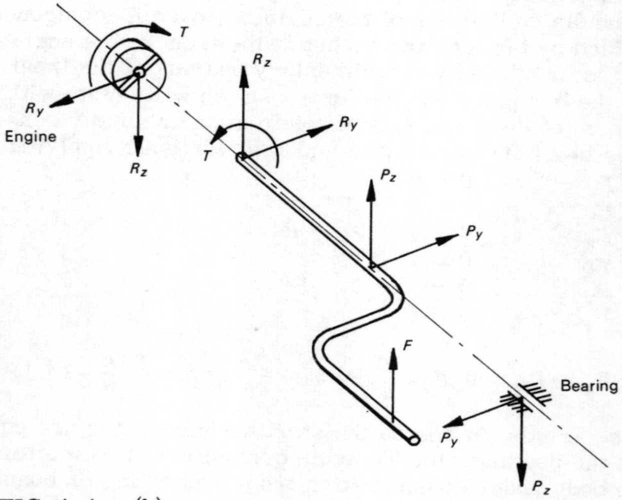

FIG. 4.4 (b)

Isolation

The idea of isolation is so important in mechanics that it deserves some discussion. As with the starting-handle, all the bodies in an assembly can be isolated from each other, so long as all the internal forces are put into the free-body diagrams.

Consider for example the plane truss in Fig. 4. 5, made up of three members fixed together by three pins. Of course the whole truss is a 'rigid body' in the statics sense, so we can start by isolating that, as in Fig. 4. 5 (b). Obviously equilibrium and symmetry demand that under a vertical load P at the apex, the

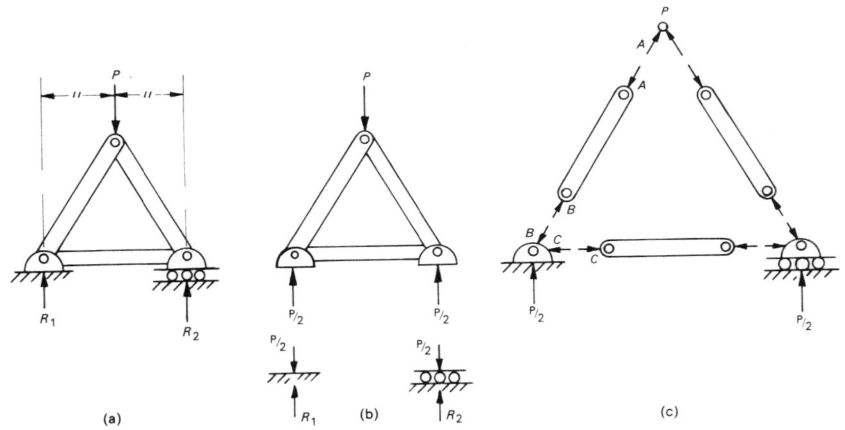

FIG. 4. 5 Isolation

reactions R_1 and R_2 should both be vertical and equal to $\frac{P}{2}$.

All the bodies in an assembly need not necessarily be isolated in this way; it is often more useful to 'dismember' the assembly only partially, as Figs. 4. 6 and 4. 7 show. It is often helpful to draw an 'isolation boundary' round that part of the assembly being considered; you must then be careful that all the relevant forces are shown. The condition of equilibrium is that all the forces crossing the boundary should themselves be in equilibrium.

It is important to remember that weight (as in Fig. 4. 6) or any other 'action at a distance', such as magnetic force, crosses any isolation boundary drawn round the body experiencing the action. This then is always true of the body's own weight, which in many structures is not negligible compared with the other forces; indeed in some, such as long-span bridges, the weights of the members of the structure are the dominant forces.

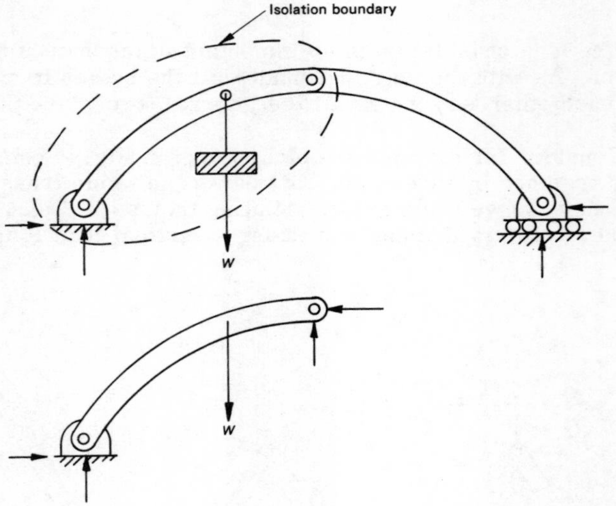

FIG. 4. 6 Three-hinge arch

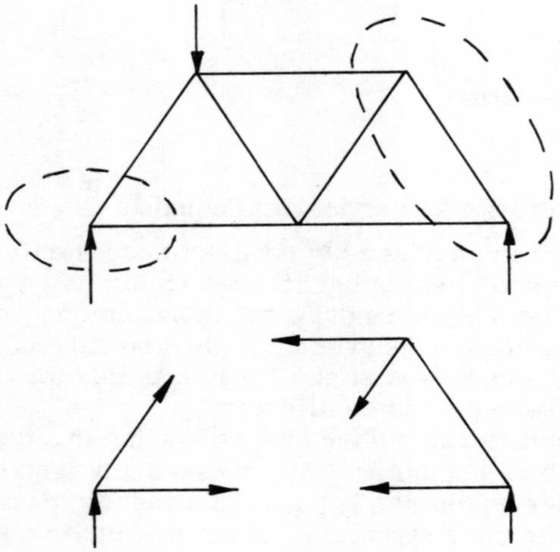

FIG. 4. 7 Isolation diagrams for a plane truss

Figure 4.7 shows isolation boundaries drawn *through* members of a plane truss. This is simply a large-scale version of what we did in Fig. 4.2 in cutting up a body into many particles. This technique is indispensable in working out how the forces are transmitted *through* a body. For example the first step in working out the stresses in the cantilever shown in Fig. 4.8 is to draw an isolation boundary which cuts the beam at distance x from the point of application of the

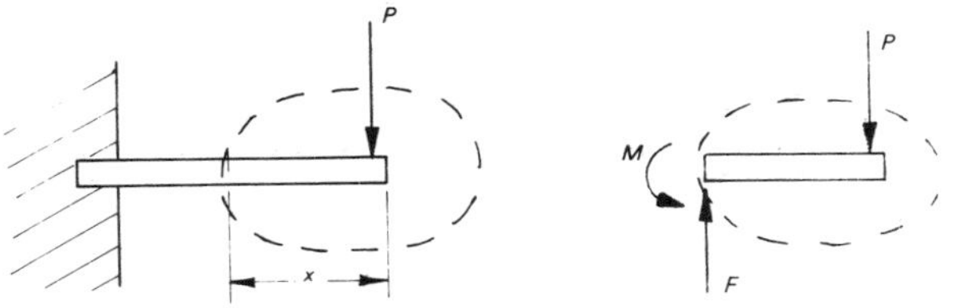

FIG. 4.8 Cantilever beam

load P. If we can neglect the weight of the beam in comparison with P, then the only external force crossing the boundary is P itself. So equilibrium obviously demands that a force F and a couple M should be transmitted across the boundary where it cuts the beam, i.e. by the part of the beam which has been removed in the isolation. This force and couple transmitted through the beam itself are called the *shear force* and *bending moment* and in this example their values are clearly

$$F = P, M = Px$$

A number of examples will be tackled in the following pages, and they will illustrate and employ the principles on the last few pages. Actually it will be seen that most difficulties are presented by trigonometry and arithmetic, rather than by the mechanics. But first we must find out what sorts of statics problems can in principle be solved by *rigid body statics*.

Limitations of rigid body statics

Transmissibility has already been mentioned. If the body is able to transmit the forces imposed upon it, the principle of transmissibility can be used in connection with the equilibrium of the *external* forces, but *not* if the *internal* transmission of force is being considered.

How stiff is rigid? The limitation of rigidity is not a severe one. If the body is not stiff enough to keep the geometry sensibly constant, then of course the deformations must be allowed for in the analysis; the subject is then deformable body statics. But an *assembly* of 'rigid' bodies may be very flexible (a mechanism) and still amenable to treatment by rigid body statics within the limitations of low (mass x accelerations)s.

Redundancy. Rigid body statics runs into trouble when there are *redundant* external constraints or internal members in an assembly of bodies. In this context 'redundant' does not mean unnecessary or even undesirable—it merely means *more than in principle is absolutely necessary to maintain equilibrium*.

Consider first a redundant constraint or support. Suppose you are one of *three* men carrying a long ladder as in Fig. 4. 9 (a). If you relax for a

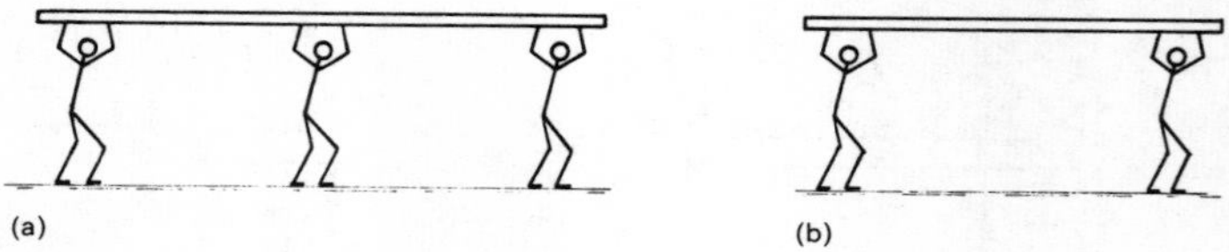

FIG. 4. 9 Men carrying ladders

moment, the ladder will not fall, whichever of the three you are (provided the other two are strong enough to support it without your help). But if you are one of *two* men carrying a ladder, as in Fig. 4. 9 (b), then of course if you let go, the thing will fall. In (a) there is a redundancy, in (b) there is not.

In more mechanical terms the situations is represented in Fig. 4.10 (a) by a rigid beam supported by three springs; a more likely structural version

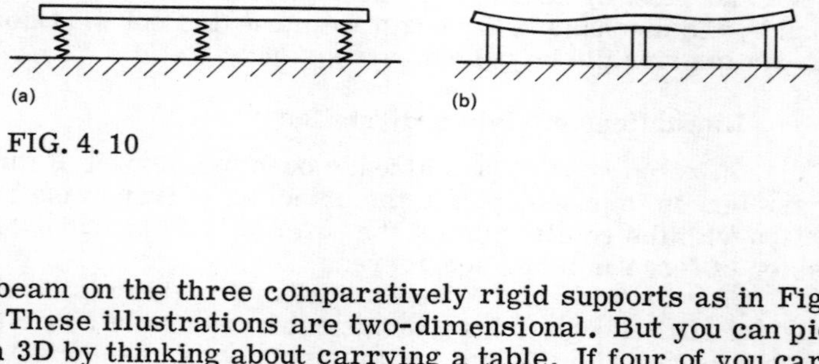

FIG. 4. 10

is a flexible beam on the three comparatively rigid supports as in Fig. 4.10 (b).

These illustrations are two-dimensional. But you can picture the same thing in 3D by thinking about carrying a table. If four of you carry a rectangular table, one of you is 'redundant'.

If you apply the conditions of equilibrium to the beam in Fig. 4.10, you will not be able to find how its weight is shared by the three springs in (a) or by the three supports in (b). The trouble is that there are *three* unknowns (the reactions at the supports) and only *two* equations of equilibrium (all the others give 0 = 0). No doubt you are already thinking that in Fig. 4. 10 (a), by the symmetry of the thing, if the beam is rigid (by comparison with the springs) then all the springs will deflect by the same amount, and so will carry equal shares of the beam's weight. Very good: but you have taken a lot for granted. You have assumed that all the springs have the same load/deflection characteristic. But what if two of the springs came from a motor-car suspension and the third from a cigarette lighter? What if two of them are made of steel and one of polythene? Think also about Fig. 4. 10 (b): will the three rigid supports carry equal shares of the flexible beam's weight?

The redundancy has brought with it two complications: the need to consider (a) deflections, and inevitably therefore, (b) the mechanical properties of the materials from which the bits and pieces are made. In more formal terms, the solution of a statics problem with a redundant constraint requires consideration of

(a) the conditions of equilibrium,
(b) the load/deflection characteristics,
(c) the compatibility of the displacements (i.e. the geometrical requirement that all the deflected and displaced members of the assembly still fit together).

The same is true of an assembly with an *internal* redundancy; it may be an additional member to minimum requirements, or it may be an internal constraint *between* members, such as a rigid joint instead of a pin joint. In Fig. 4. 11 (a) the members AB and BC are made continuous by a rigid joint. The

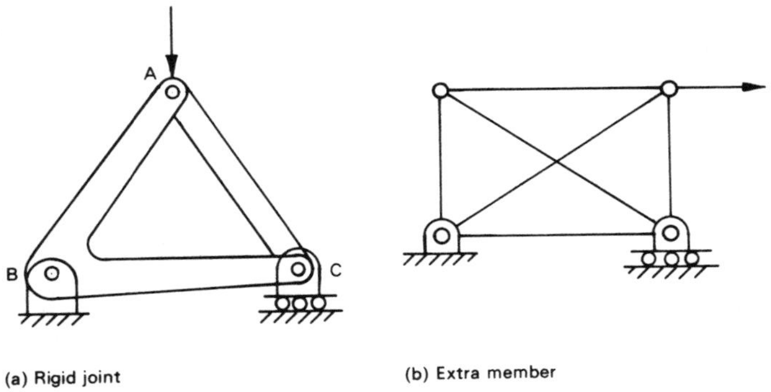

FIG. 4. 11 Structures with 'redundancies'

member AC could be removed, if ABC were strong enough, although you may not think that it would be a very sensible thing to do.

Finally, note that these 'redundancies' are only redundant in the strict statical sense; they may be sensible and desirable, technically and otherwise (if the ladder is very long it may be too *heavy* for two men to carry). Structural engineers usually call structures with redundancies either 'statically indeterminate' or 'hyperstatic' structures. Anyone wishing to pursue the structural aspects will find these principles and the criteria for determinacy, and many examples, set out clearly by Brown [3], or by Marshall and Nelson [4]. The statically determinate structure is on the dividing line between the hyperstatic structure and the mechanism or machine (a structure that 'goes').

Summarising, rigid body statics alone is helpless in analysing (a) redundant or hyperstatic structures, but can answer many questions in (b) simple stiff structures and (c) low-acceleration machines and flexible structures.

Some examples of rigid body statics

The *six* conditions of equilibrium (4.2) are reduced in number for some special cases:

(a) to *three* if all the forces are parallel,
(b) to *three* if all the forces act in a single plane (a coplanar system),
(c) to *two* if all the forces are parallel and coplanar.

Two further special cases are:

(d) if only *two* forces act on a body, they must be equal and opposite and in the same line,
(e) if only *three* forces act on a body, they must be coplanar and intersect at a single point (i.e. be concurrent).

Using the general conditions of equilibrium, satisfy yourself that these simplifications are correct.

Single bodies

A chestnut amonst problems on equilibrium is the garden wheelbarrow lifted with only one hand. Anyone who has tried it will testify to its impossibility, but it is useful as an example of taking components of a moment. Figure 4.12 shows a garden barrow of weight W, whose shafts (i.e. handles) are near-enough in a horizontal plane and make angles α with the centre line. The relevant dimensions are shown on the diagram. The problem is to find the magnitudes of **F** and **M** needed to lift the two legs just clear of the ground. **F** and **M** are shown in the directions they must clearly have.

The coordinate axes have their origin at the centre of the wheel and the x axis on the centre line of the barrow. The moment of the couple is clockwise

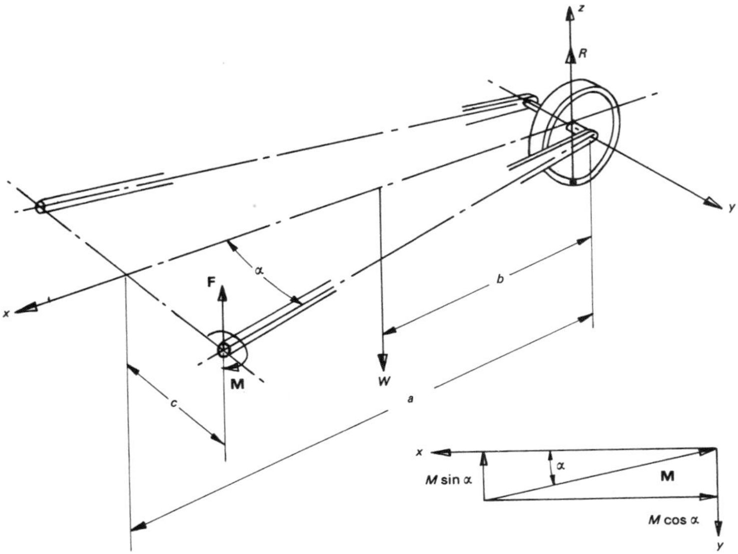

FIG. 4. 12 Wheelbarrow

therefore as:

looking along the handle towards the wheel, so both the x and y components are negative. The equations of equilibrium are:

$$
\begin{array}{ll}
F_x & 0 = 0 \\
F_y & 0 = 0 \\
F_z & F + R - W = 0 \\
M_x & -M\cos\alpha + Fc = 0 \\
M_y & -M\sin\alpha + Wb \\
M_z & 0 = 0
\end{array}
$$

These can easily be solved to yield M and F.

Turning now to another problem, Fig. 4.13 represents an arrester gear in the form of a circular ring which 'catches' hooks carrying loads in all directions. The designer wants to know the bending and twisting moments, etc., at the point where the ring stalk is held, i.e. the origin of coordinates in the diagram. The problem therefore is to find the moments and forces at the origin due to a force in an arbitrary direction applied at any point on the ring.

Suppose a force **P** is applied to the ring at a point making angle θ with the x axis and in a direction ($X\mathbf{i} + Y\mathbf{j} + Z\mathbf{k}$). The force **P** can be written

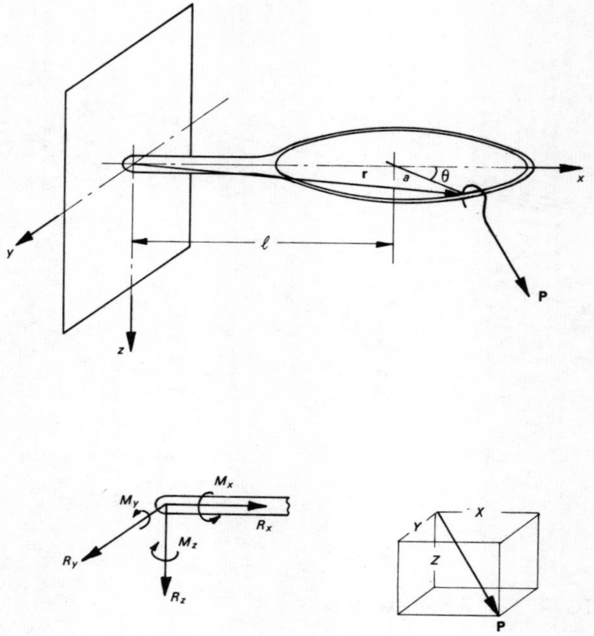

FIG. 4. 13 Arrester gear

$$\frac{P}{\sqrt{X^2 + Y^2 + Z^2}} (X\mathbf{i} + Y\mathbf{j} + Z\mathbf{k}) = P_x\mathbf{i} + P_y\mathbf{j} + P_z\mathbf{k}, \text{ say}$$

(satisfy yourself that **P** *can* be so written).

Isolating the ring, we see that it is in equilibrium under **P** and the forces and moments at the origin, shown in the inset in Fig. 4. 13.

R_x is a tension or compression, R_y and R_z are components of the shear force; M_x is a twisting moment or torque, and M_y and M_z are the components of the bending moment. Applying (4. 1) we obtain directly

$$\mathbf{R} + \mathbf{P} = 0$$

or, resolving in the coordinate directions,

$$R_x + P_x = 0, \text{ etc.}$$

and $$\mathbf{M} + \mathbf{r} \wedge \mathbf{P} = 0$$

or $$(M_x\mathbf{i} + M_y\mathbf{j} + M_z\mathbf{k}) + \mathbf{r} \wedge \mathbf{P} = 0$$

i.e.
$$\mathbf{r} \wedge \mathbf{P} = \begin{vmatrix} \mathbf{i} & \mathbf{j} & \mathbf{k} \\ x & y & z \\ P_x & P_y & P_z \end{vmatrix}$$

$\mathbf{r}$ is the point of application of $\mathbf{P}$ and is

$$\mathbf{r} = x\mathbf{i} + y\mathbf{j} + z\mathbf{k}$$

where $x = l + a \cos \theta, y = a \sin \theta, z = 0$

So the forces and moments at the origin are:

Compressive force, $R_x = -P_x$

Shear force, $(R_y\mathbf{j} + R_z\mathbf{k}) = -(P_y\mathbf{j} + P_z\mathbf{k})$

Torque, $M_x = -(yP_z - zP_y) = -yP_z$

Bending moment, $(M_y\mathbf{j} + M_z\mathbf{k}) = -(zP_x - xP_z)\mathbf{j} - (xP_y - yP_x)\mathbf{k}$

which is probably as far as it is worth taking the algebra.

Balancing of rotors

A final 'single-body' application which is worth mentioning is the problem of balancing a rotor which consists of a number of masses fixed together and rotating as one. It is exemplified by a turbine rotor consisting of a number of wheels mounted on one shaft. In general the centres of mass of the individual wheels will not be exactly on the geometrical axis of the shaft. Furthermore it is not a practical proposition to make them so. Now when the rotor rotates at high speed, each wheel will require from the shaft a radial force of $me\Omega^2$ to keep the centre of mass moving in its small circle of radius e. The eccentricity e of the centre of mass will be extremely small in an accurately manufactured turbine wheel, but Ω^2 will be very large indeed in a modern turbine.

All these radial forces must be provided ultimately by the bearings carrying the rotor, except insofar as they balance each other out. If such balance is achieved, the only forces on the bearing are those required to support the weight of the rotor. The problem therefore is to find the conditions for such a balance, to find for instance what additional masses, and where, will bring balance to an out-of-balance rotor. It is not sufficient that the centre of mass of the whole assembly is on the axis, although that is a necessary condition (and will, incidentally, assure *static* balance); you can check this by spinning the cranks of a bicycle with the chain off, when you will see an out-of-balance rotating couple rocking the bicycle frame.

The situation is shown in Fig. 4.14. $\mathbf{P}_1$, etc., are the forces sustained by the shaft in keeping the centres of mass rotating in small circles of radius e_1, etc.:

$$\mathbf{P}_1 = m_1 e_1 \Omega^2 \hat{\mathbf{e}}_1$$

where $\hat{\mathbf{e}}_1$ is a unit vector.

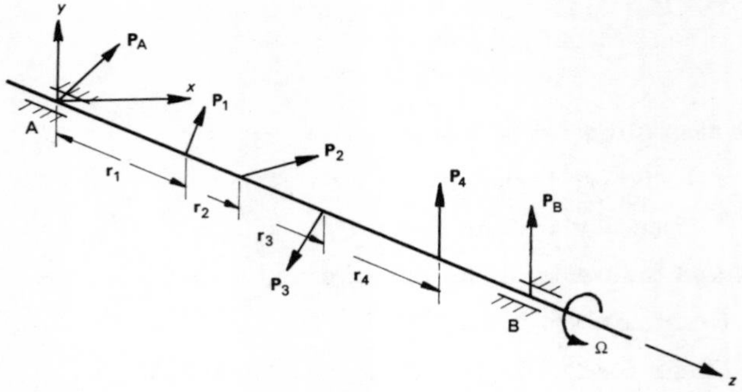

FIG. 4. 14 Rotor

Plainly all the forces are rotating at the same speed Ω about the z axis, so the rotating force picture can be treated as a 'static' problem.

The equations of equilibrium will straightforwardly give the forces at the bearings for an out-of-balance rotor, or more usefully, can be used to give the conditions necessary to make them zero (apart from the weight of the rotor); i.e. to balance the rotor.

The conditions of equilibrium are

$$\Sigma \mathbf{P} = 0, \quad \Sigma \mathbf{r} \wedge \mathbf{P} = 0$$

All the **r**'s are in the z direction, and all the **P**'s are rotating in planes parallel to the xy plane (i.e. perpendicular to the z axis), so all the $\mathbf{r} \wedge \mathbf{P}$'s are in the xy plane and perpendicular to the corresponding **P**'s.

To make $\mathbf{P}_A$ and $\mathbf{P}_B$ both zero, in general requires the addition of *two* further masses to the rotor. Two additional masses introduce six new quantities—mass, orientation and distance along the axis, for each mass—whereas there are only four unknowns in the equilibrium equations. So there is freedom in deciding two of the quantities; e.g. the axial positions can be fixed and then the masses and orientations calculated.

The equilibrium conditions can best be displayed by drawing vector diagrams—a couple polygon and a force polygon.

Simple stiff structures

This is not the place for any elaborate structural analysis, so the examples below are simple ones on structures with members which can carry axial loads only, i.e. members in simple tension or compression only, connected together by pin joints. Real structures are never made with pin joints (well hardly ever), but for analysis it is often in order to assume that they are.

Figure 4.15 shows a simple space frame for supporting an overhanging load. We can find the forces in the three members which meet at A by taking the origin at A and isolating the joint at A. Denoting the forces in the

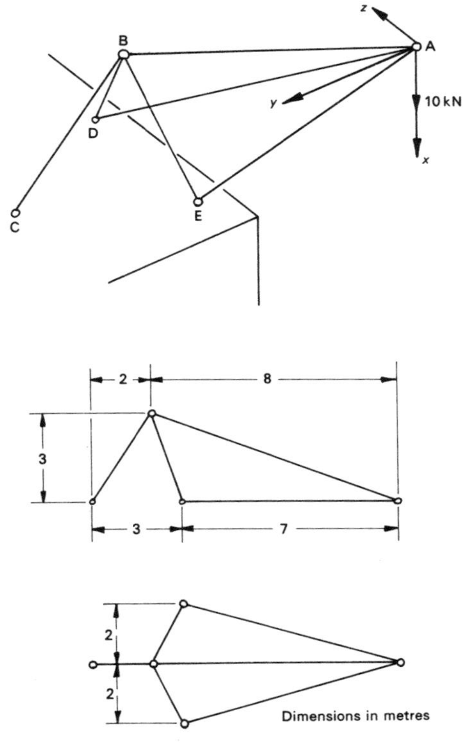

FIG. 4.15 Space frame

members by $\widehat{AB}$, $\widehat{AD}$ and $\widehat{AE}$, we call them all tensile (positive) and then know that a negative force will be compressive. With the origin at A, the other ends of the members are at:

$$\begin{aligned}
&\text{B:}\quad x = -3, \quad y = 8, \quad z = 0\\
&\text{D:}\quad x = 0, \quad y = 7, \quad z = 2\\
&\text{E:}\quad x = 0, \quad y = 7, \quad z = -2
\end{aligned}$$

Since all the forces on joint A intersect, there are no moment equations of equilibrium, just forces.

$$\Sigma F = 0 = 10i + \frac{\overset{\frown}{AB}}{\sqrt{3^2 + 8^2}}(-3i + 8j)$$

$$+ \frac{\overset{\frown}{AD}}{\sqrt{7^2 + 2^2}}(7j + 2k) + \frac{\overset{\frown}{AE}}{\sqrt{7^2 + 2^2}}(7j - 2k)$$

$$\sqrt{3^2 + 8^2} = 8.54$$

$$\sqrt{7^2 + 2^2} = 7.28$$

$$x(i):\ \ 0 = 10 - \frac{3}{8.54}\overset{\frown}{AB}$$

$$y(j):\ \ 0 = \frac{8}{8.54}\overset{\frown}{AB} + \frac{7}{7.28}\overset{\frown}{AB} + \frac{7}{7.28}\overset{\frown}{AE}$$

$$z(k):\ \ 0 = \frac{2}{7.28}\overset{\frown}{AD} - \frac{2}{7.28}\overset{\frown}{AE}$$

These yield directly

$$\overset{\frown}{AB} = +\ 28.5\ \text{kN, tension}$$

$$\overset{\frown}{AD} = \overset{\frown}{AE} = -\ 13.9\ \text{kN, compression}$$

An oblique force such as $(10i - 6j + 3k)$kN would present no additional complication; neither would lack of symmetry in the frame, just a bit of extra arithmetic.

The systematic isolation of all the joints in turn is known as the 'method of joints'. Clearly in a frame with a lot of members, there is a proliferation of simple calculations. For space frames, a tabular method employing Southwell's 'tension coefficients' is commonly used. For plane trusses, the same process can be used, but often, in cases of complicated geometry, it is much easier and quicker to use a graphical method based on Maxwell [5], labelling the structure according to 'Bow's notation'. Much fuller details can be found in any book on elementary structural analysis, such as that by Brown [3], from which the last example was taken.

Whatever the details, the method of joints should be started at a joint where there are no more unknowns than equations of equilibrium. Since the joint is only acted upon by intersecting forces (no moment equations), this means no more than two unknowns in a plane truss, and no more than three in a space frame. For example in the truss shown in Fig. 4.16, once R_1 and R_2 have been calculated by overall equilibrium of the whole truss, a start could be made at joint A, where there are two unknown member forces, but not at joint B, where there are three. It would be worth spending a few minutes putting in numbers for

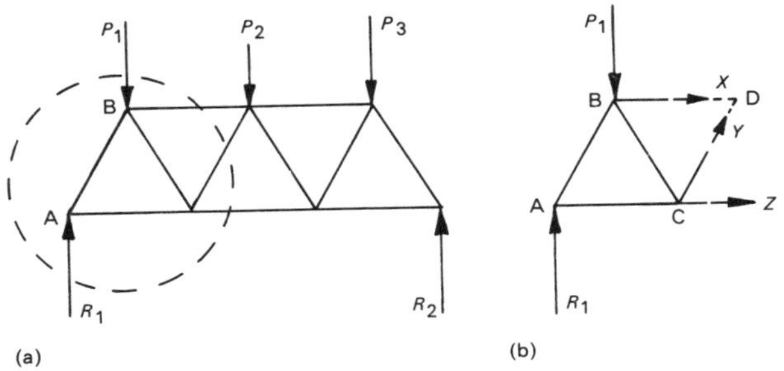

FIG. 4.16

P_1, P_2 and P_3 and working out all the member forces by the method of joints. Note that all the members have the same length.

The dotted curve in Fig. 4.16(a) is there to remind us that we can draw an isolation boundary anywhere, so long as we include in the equilibrium considerations all the forces that cross the boundary. Figure 4.16 (b) shows the part of the truss isolated, and all the forces acting on it, assuming that the weights of the members are small compared with the applied forces, which is usually the case in small trusses.

Since all the forces do not intersect we shall now have at least one moment equation. There are three unknowns, $\boldsymbol{X}$, Y and Z, so we need all the three equilibrium equations for the general plane case. We can either resolve forces in *two* directions and take moments about *one* point, or resolve forces in one direction and take moments about two points. Indeed we could take moments about three points so long as they were not in a straight line. All these approaches would yield the same information ultimately.

In this particular case there is scope for the old dodge of taking moments about the point where two unknown forces intersect. Work out the forces $\boldsymbol{X}$, Y, Z for your own example, by (a) taking moments about C, (b) taking moments about D, and (c) resolving forces either horizontally or vertically.

Drawing general isolation boundaries like the one just used makes it possible to find the forces in just a few members anywhere in the frame, without working through the whole frame. In structures this is known as Ritter's 'method of sections'. It is useful when the method of joints 'gets stuck' by not being able to proceed to a joint with few enough unknowns to be soluble.

Flexible structures

Remembering our definition of static (negligible mass x accelerations's), we can apply statics methods to flexible structures and mechanisms so long as we allow for the changes of *geometry*. A few examples idealised to two dimensions will make the point.

Figure 4.17 (a) shows the lifting end of a garage jack which raises a load *P* on the platform CD by extending the link AC hydraulically. To find the forces in the links BC, AC and AD for a given *P*, we isolate the platform and use the conditions of equilibrium for the four coplanar forces shown in Fig. 4.17 (b). Note that all the links can carry axial loads only.

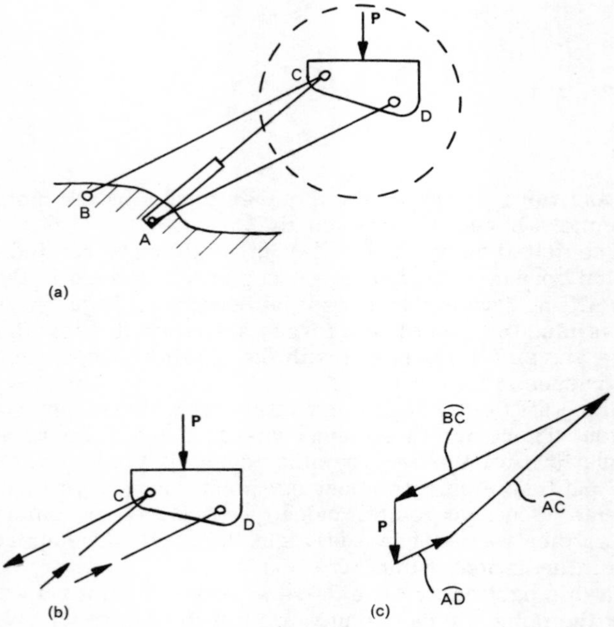

FIG. 4.17 Garage jack

First we remind ourselves of the steps in establishing the conditions of equilibrium: we transfer all the force to an origin of our own choice, introducing appropriate couples; then the sum of all the forces must be zero and the sum of all the couples must be zero. A sensible point to take as origin in Fig. 4.17 (b) is the point C. Then the couples to be introduced are equal to the moments of **P** about C, and of the force in AD about C. Their sum must be zero, so the force in AD can be found. Finally the sum of *four* forces acting at C must be zero; two of them are known, **P** and $\widehat{AD}$, so the other two can be found by simple trigonometry or by a graphical construction of the polygon of forces as in Fig. 4.17 (c).

The front suspension of the author's rather staid old motor car amounts in essence to the mechanism shown in Fig. 4.18. AB and BD are 'two-force members' and so the forces they transmit must act along their centre-lines

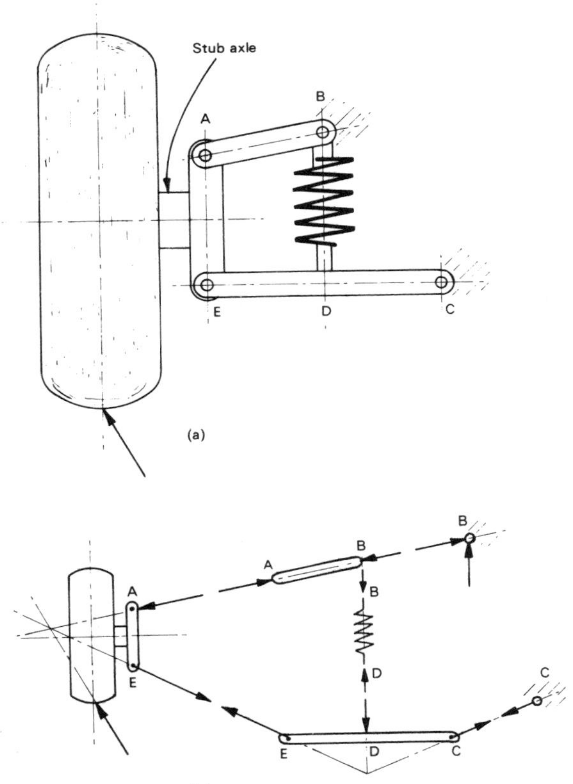

FIG. 4.18 Suspension

(the lines joining the pins at the ends). The link EDC, and the body consisting of the wheel, the stub axle and the link AE, are both 'three-force members' and so in each case the three forces must be concurrent. All the bodies are isolated in Fig. 4.18 (b), and the forces acting on each body shown. For any given tyre force the whole thing can easily be solved, either numerically or graphically.

Figure 4.19 is a sketch of an excavator. The designer's problems would include finding the forces in the cables, and in the parts of the mechanism for various orientations of the assembly and values of the bucket force **F**. Isolation (1) in Fig. 4.19 (b) shows how the tension T_1 and the force at B could be found by considering the equilibrium of the three force member. In isolation (2) the tension T_2 could first be found by taking moments about A, and then the force at A would be the only unknown. Finally amongst these examples is the 'lazy-tong'

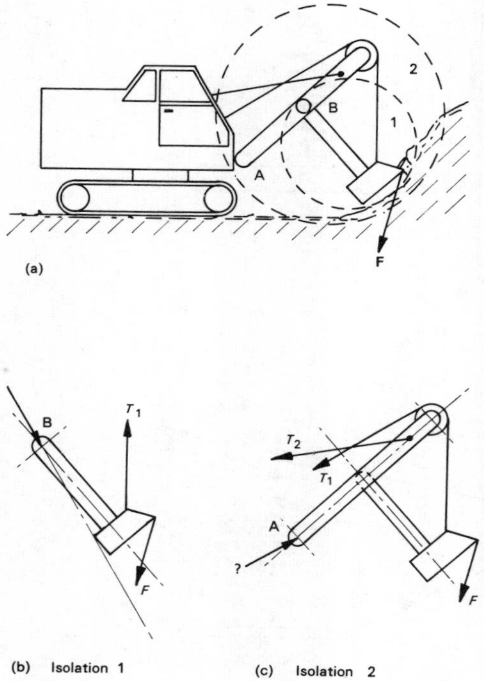

FIG. 4.19 Excavator

lifting grab, Fig. 4.20, the sort of example used by most books on this subject to drive the reader into the waiting arms of an energy method of solution of the sort introduced in the next paragraph. Make a few sketches to indicate how you would find the forces by isolation of the members. You might then be only too happy to read the next few pages on work and energy.

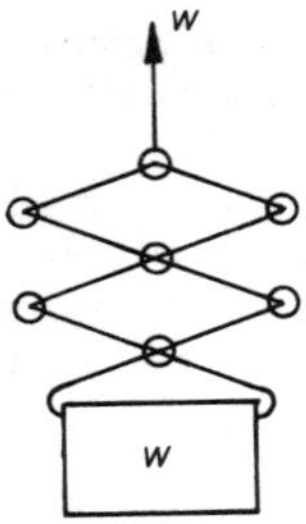

FIG. 4. 20 Lazy-tongs

Work and Energy

'Every general principle involves an economy of thought'.

Ernst Mach

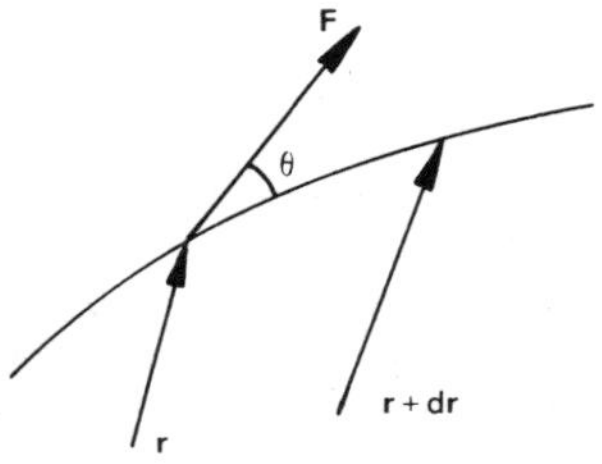

FIG. 4. 21

We shall be learning quite a lot about work and energy in Chapter 6, but for the present we simply define the work done by a force, when its point of application moves from A to B, as the scalar quantity

$$W = \int_{\mathbf{A}}^{\mathrm{B}} \mathbf{F} \, . \, \mathbf{dr}$$

or

$$W = \int_{\mathrm{A}}^{\mathrm{B}} F_x \, \mathrm{d}x + \int_{\mathbf{A}}^{\mathrm{B}} F_y \, \mathrm{d}y + \int_{\mathbf{A}}^{\mathrm{B}} F_z \, \mathrm{d}z$$

$$\mathbf{F} \, . \, \mathbf{dr} = F \cos \theta \, \mathrm{d}s$$

where θ is the angle between **F** and the path of its point of application, and ds is the increment of length of the path between **r** and **r** + **dr**; see Fig. 4. 21, and the definition of work in Chapter 2.

Plainly, for any individual force, W may be positive or negative according to whether the point of application moves generally 'with' or 'against' the direction of the force. It is important to appreciate that no work is done by a force through a displacement which is perpendicular to the force.

Remember that work is a scalar quantity, with dimensions of force x length (ML^2/T^2) and is expressed in 'joules' in the SI system.

A simple example is the work done by a force in compressing a spring, Fig. 4. 22.

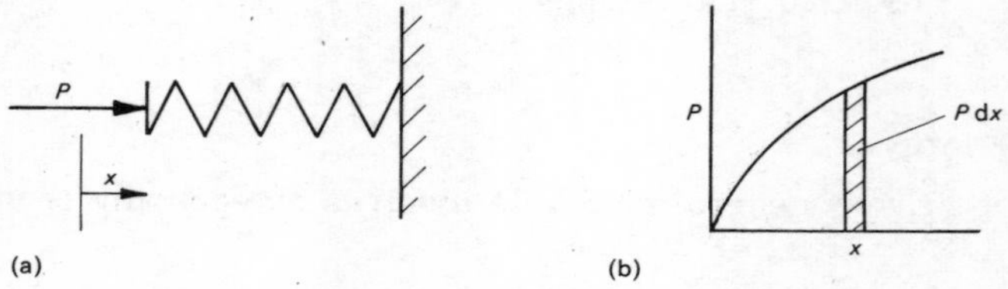

FIG. 4. 22

$$W = \int P \, dx$$

and so clearly the work done is equal to the area under the P/x curve. If the spring is an elastic spring (i.e. the P/x curve is reversible), the work done by the force is stored in the spring and can be recovered. The energy stored is called *elastic strain energy*.

If the spring is linear (i.e. $P = kx$, where k is called the *stiffness*), then

$$W = \tfrac{1}{2}Px = \tfrac{1}{2}kx^2 = \frac{P^2}{2k}$$

All real bodies are flexible to some extent, so these expressions apply to all linear elastic bodies. For such a body, or an assembly of such bodies, *the total work done by the external forces equals the change in the elastic strain energy stored.* (This omits gravitational potential energy because at this stage we are only concerned with small displacements and with bodies whose weights are much smaller than the external forces; we shall come to grips with potential energy in Chapter 6.)

In the limiting case of an assembly of rigid bodies, the total work done by the external forces is therefore zero—a rigid body cannot absorb any strain energy: for a rigid structure all the individual work terms due to the external forces must be zero because there can be no displacements; for a mechanism or machine made up of rigid links, the *total* work is zero, the 'work in' being equal to the 'work out', neglecting friction and transfer of energy by heat.

Effectively intermediate between 'all-elastic' and 'all-rigid' is a set-up like that in Fig. 4.18 where practically all the flexibility is vested in two of the members, the tyre and the spring, the other bodies being rigid by comparison.

Note in all this that work is an action, essentially associated with *motion*, whereas energy is a property of the *state* of a body or system of bodies.

There are several work methods in mechanics, and we are only scratching the surface of a collection of powerful techniques, whose applications you will eventually encounter in stress analysis, structural analysis, vibration and dynamics. For our purposes, where we may be concerned with substantial changes of geometry, the most valuable is that now known as 'virtual work'. The next few paragraphs take only an elementary look at the principle of virtual work, but modern developments have given it a central position in structural analysis.

Virtual work

The use of a work method in equilibrium situations (where there is generally *no motion*) goes back to the work of Stevinus on pulleys and of Galileo on weights on inclined planes. The universal validity of their principle in all cases of equilibrium was recognised in 1717 by John Bernoulli*, who introduced the term 'virtual velocities'. That term is usually replaced in modern writings by 'virtual displacements', but both concepts are useful.

The principle of virtual work is

In an assembly of bodies in equilibrium, the total work done on the assembly by the external forces in any small, geometrically compatible displacements is equal to the total energy stored or dissipated within the assembly.

A few points need elaboration:

If there is no friction, there will be no energy *dissipated*. The virtual work method loses much of its simplicity if friction has to be taken into account.

If all the bodies in the assembly are rigid, there will be no energy *stored*.

Provided the displacements postulated are geometrically compatible with each other, and do not violate the geometrical properties of the bodies in the assembly, they need not be the *actual* displacements. They may be a fictitious set of displacements. There may in reality be no displacements at all; hence the term 'virtual displacements'.

Small displacements are specified so that the forces and geometry *remain constant* to the first order.

The proof of the principle is self-evident for a particle and can be extended to a body and to an assembly of bodies. The work done in any small displacement by all the forces acting on a particle in equilibrium is zero, because all the forces have a zero resultant, which will do zero work in a small displacement.

* See [6]

Consider a few simple examples. Figure 4.23 (a) shows two (frictionless) sliders connected by a rigid link. What will be the relation between the external forces P and Q for equilibrium? Solving by resolving the forces as in

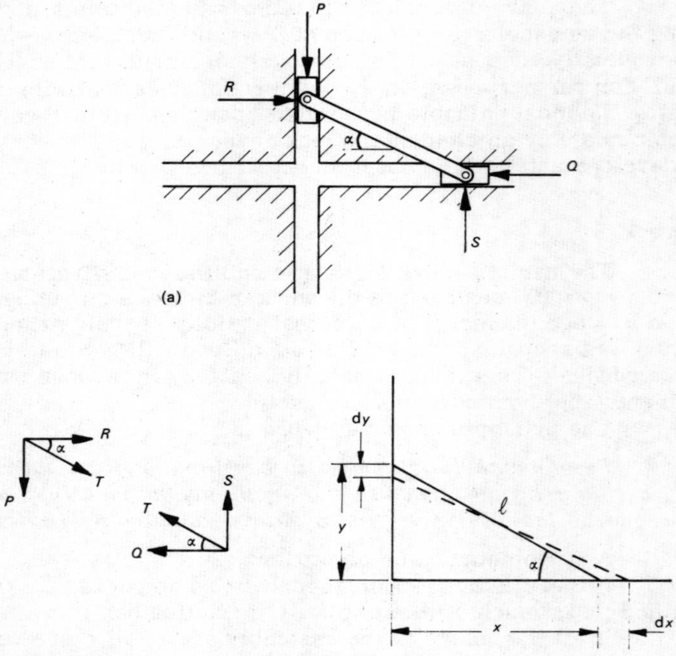

(b) Resolution (c) Virtual work

FIG. 4.23

Fig. 4.23 (b) gives straightforwardly

$$P = T \sin \alpha, \quad Q = T \cos \alpha, \quad \text{i.e.} \frac{P}{Q} = \tan \alpha$$

To do it by virtual work we contemplate an infinitesimal displacement as shown in (c). Having regard to sign (P and Q are both shown acting in directions opposite to positive dy and dx), virtual work gives

$$-P\,dy - Q\,dx = 0$$

$$\frac{P}{Q} = -\frac{dx}{dy}$$

Note that R and S have no part in the work equation. Geometry gives

$$y = l \sin\alpha : \quad dy = l \cos\alpha \, d\alpha$$
$$x = l \cos\alpha : \quad dx = -l \sin\alpha \, d\alpha$$

which give

$$\frac{P}{Q} = \tan\alpha$$

In this single-body example, the work method affords no simplification at all. It is in connected assemblies that it comes into its own.

The lazy-tongs of Fig. 4.20 are shown again in Fig. 4.24. When

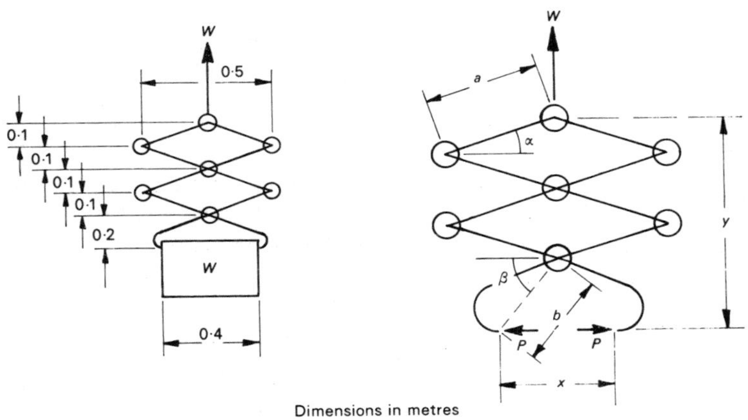

FIG. 4.24 Lazy tongs

lifting the block of weight W, suppose it takes up the configuration with the dimensions shown. What will the gripping force be? To apply virtual work, replace the block by the two forces P distance x apart, when the tongs have height y. For virtual displacements dx and dy the work equation is

$$P dx + W dy = 0$$

Both signs are positive because P and W are both acting in the directions of positive dx and dy.

So $$P = -W \frac{dy}{dx}$$

That is the mechanics. The rest of the story is geometry.

$$y = b \sin\beta + 4a \sin\alpha$$
$$x = 2b \cos\beta$$
$$dy = b \cos\beta\, d\beta + 4a \cos\alpha\, d\alpha$$
$$dx = -2b \sin\beta\, d\beta$$

Because the links are taken to be rigid,

$$d\alpha = d\beta$$

so $$\frac{dy}{dx} = -\frac{b \cos\beta + 4a \cos\alpha}{2b \sin\beta}$$

For the dimensions shown,

$$b \cos\beta = 0.2 \text{ m}$$
$$a \cos\alpha = 0.25 \text{ m}$$
$$b \sin\beta = 0.2 \text{ m}$$

so $$\frac{dy}{dx} = -\frac{0.2 + 1.0}{0.4} = -3$$

$P = 3W$ is the gripping force.

Figure 4.25 shows an example of a flexible structure, a pantograph mechanism. The two circles represent equal gear wheels which are in mesh, and so must rotate with equal and opposite angular velocities and displacements. The

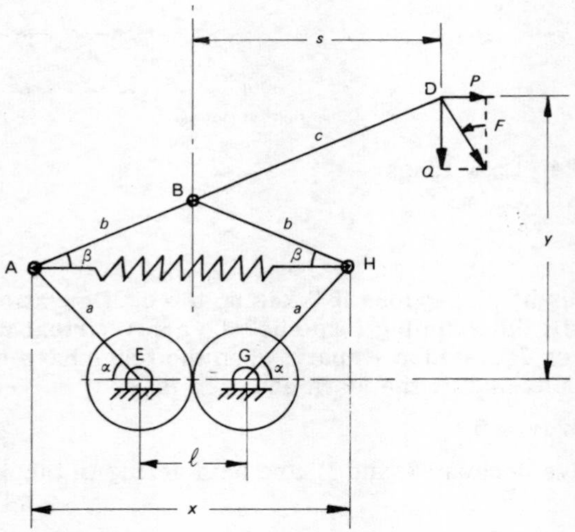

FIG. 4.25 Pantograph

arms EA and GH are fixed to the gear wheels and must therefore rotate with them; the assembly is symmetrical up to the point B, as shown by the various lengths and angles.

We set ourselves the task of finding the necessary spring tension T to hold the mechanism in equilibrium in the position shown under the applied force F at the end of the rigid arm ABD. The force F has horizontal and vertical components P and Q as shown. Before we tackle it by virtual work, you might like to contemplate the labour of finding the solution by isolation and drawing free body diagrams; the three-force body ABD, the two-force bodies AH and BH, the bending moments and forces in the arms AE and GH, the forces between the gear teeth, and the reactions at the supports.

In the work method the only forces that need be taken into account are (a) the only work-doing external force F, and (b) the tension T in the spring, which is the only energy-absorbing member in the mechanism.

For the equilibrium position of Fig. 4.25, a virtual displacement ds, dy of the end of the arm D will be accompanied by a virtual change in the length of the spring dx. (Note that by the symmetry of the mechanism, pivot B can only move vertically). The principle of virtual work gives

$$P ds - Q dy = T dx$$

Make sure that you agree with the signs. The remainder of the analysis is trigonometry.

$$x = l + 2a \cos \alpha$$
$$dx = -2a \sin \alpha \, d\alpha$$

$$y = a \sin \alpha + (b + c) \sin \beta$$
$$dy = a \cos \alpha \, d\alpha + (b + c) \cos \beta \, d\beta$$

$$s = c \cos \beta$$
$$ds = -c \sin \beta \, d\beta$$

We want to be rid of $d\beta$. A second way of writing x is

$$x = 2b \cos \beta$$
$$dx = -2b \sin \beta \, d\beta = -2a \sin \alpha \, d\alpha$$

So $$d\beta = \frac{a \sin \alpha}{b \sin \beta} \, d\alpha$$

Putting the differentials in the work equation,

$$-Pc \sin \beta \frac{a \sin \alpha}{b \sin \beta} \, d\alpha - Q \left[a \cos \alpha + (b + c) \cos \beta \frac{a \sin \alpha}{b \sin \beta} \right] d\alpha$$

$$= -T \, 2a \sin \alpha \, d\alpha$$

which gives

$$T = \frac{Pc}{2b} + Q \left[\cot \alpha + \left(\frac{b + c}{b} \right) \cot \beta \right]$$

Note the very simple result for a horizontal applied force, i.e. $Q = 0$. What do the cot terms in the bracket suggest to you about the load/displacement characteristics of the mechanism?

Note especially that we have been considering the work done in infinitesimal displacements about an equilibrium position, with T and F constant. The virtual work method has taken the non-linear force/displacement characteristics in its stride.

Finally we consider the engine mechanism, but simplified by the absence of friction and significant (mass x acceleration)'s.

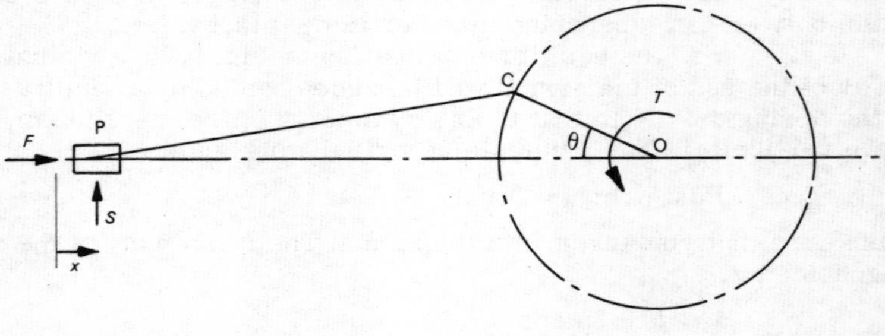

FIG. 4. 26 The engine mechanism

The work-doing forces acting on the mechanism shown in Fig. 4. 26 are the gas-force F on the piston P, and the resisting torque T on the crank, provided by whatever the engine may be driving. In the absence of acceleration effects, the mechanism, although in motion, can be treated as being in equilibrium in any instantaneous position. Applying the principle of virtual work to an infinitesimal displacement dx of the piston and the corresponding rotation dθ of the crank

$$F\mathrm{d}x - T\mathrm{d}\theta = 0$$

(The work done by a moment M through a small angular displacement $\delta\theta$ about the same axis as the moment is $M\,\delta\theta$.)

Dividing by an arbitrary time increment

$$F\frac{\mathrm{d}x}{\mathrm{d}t} - T\frac{\mathrm{d}\theta}{\mathrm{d}t} = 0$$

$\frac{\mathrm{d}x}{\mathrm{d}t}$ and $\frac{\mathrm{d}\theta}{\mathrm{d}t}$ are *any* geometrically compatible velocities. They could for instance be 'virtual velocities' in a stationary mechanism, or in the present example they could be the *actual* piston velocity v_p and crank angular velocity ω.

$$Fv_p - T\omega = 0$$

or

$$T = F\frac{v_p}{\omega}$$

We shall see in the next chapter how to find velocities in mechanisms.

In this chapter we have seen how to find the forces transmitted by stationary or slowly moving assemblies of bodies subjected to external loading.

Bibliography

1. Hill. R. *Principles of Dynamics*, Pergamon, 1964
2. Rutherford, D. E. *Vector Methods*, Oliver and Boyd, 1957
3. Brown, E. H. *Structural Analysis*, Vol. 1, Longman, 1967
4. Marshall, W. T., and Nelson, H. N. *Structures*, Pitmans, 1969
5. Maxwell, J. C., 'Reciprocal Figures and Diagrams of Forces', *Phil. Mag.*, April 1864
6. Mach, E. *The Science of Mechanics*, 1883, published in English by the Open Court Publishing Co., 1902.

Crandall, S. H. and Dahl, N. C. *An Introduction to the Mechanics of Solids*, McGraw-Hill, 1959.

Examples

Ex. 4.1 Determine the forces in all the members of the frame. Treat all the

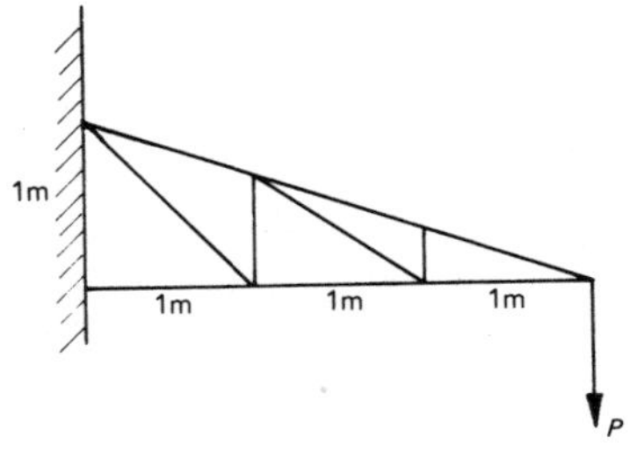

Ex. 4.1

joints as pin joints. Solve by inspection in about three minutes.

Ex. 4. 2 The members of the tube framed structure in the diagram are so

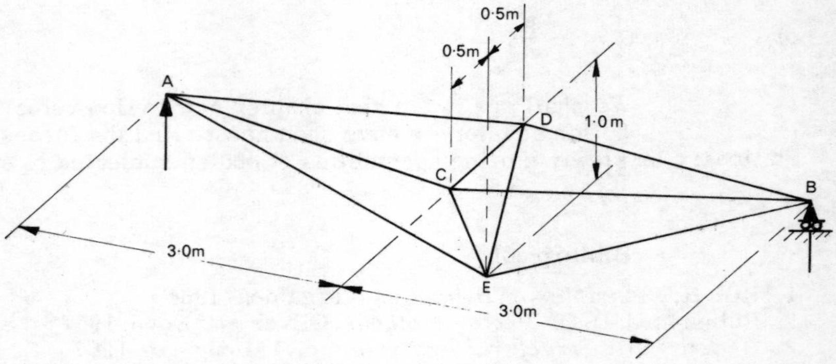

Ex. 4. 2

connected at their ends that they may be assumed pin jointed.

Calculate the forces in all the members when a vertical load P is applied at E.

Ex. 4. 3 A plane 45° truss carries the loading shown (loads in kN). Find the

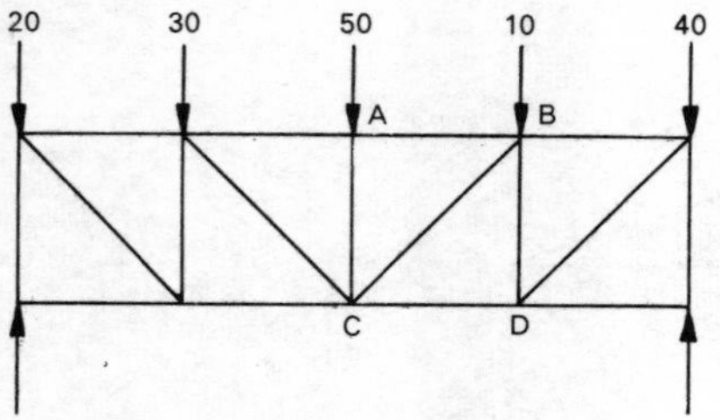

Ex. 4. 3

forces in the members, AB, CB, CD. [−70, + 30 $\sqrt{2}$, + 40 kN]

Ex. 4. 4 In the figure, dimensions are in metres and loads in kN. Calculate the

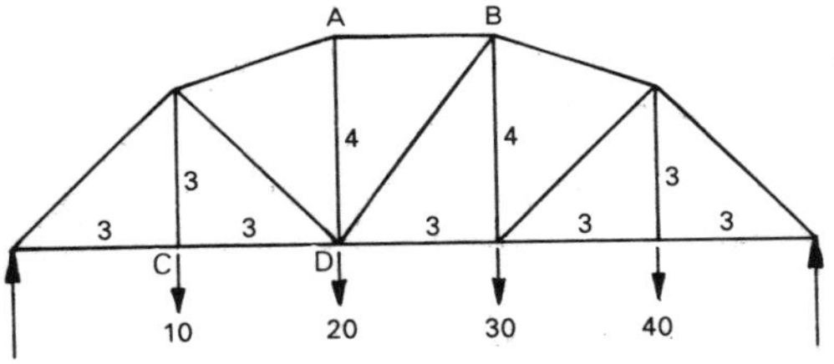

Ex. 4. 4

forces in the members AB, CD. [−52-5, + 40 kN]

Ex. 4. 5 A stiff beam AB is supported by three bars ED, DB and CB, all pin jointed. The cross-sectional area of all three bars is 10^{-3} m^2 and they are made of steel which has a Young's modulus of 2×10^{11} N/m^2.

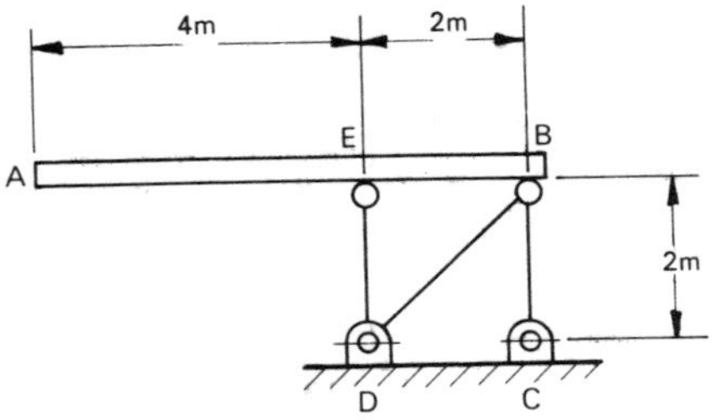

Ex. 4. 5

Calculate the forces in the three bars, and the horizontal and vertical movement of point A, if a vertically downward force of 20 kN is applied at A. Treat the beam AB as rigid. [−60, 0, 40kN, 0. 4 mm left, 2. 6 mm down]

Ex. 4. 6 In the mechanism shown wheels A and B are geared together with a ratio

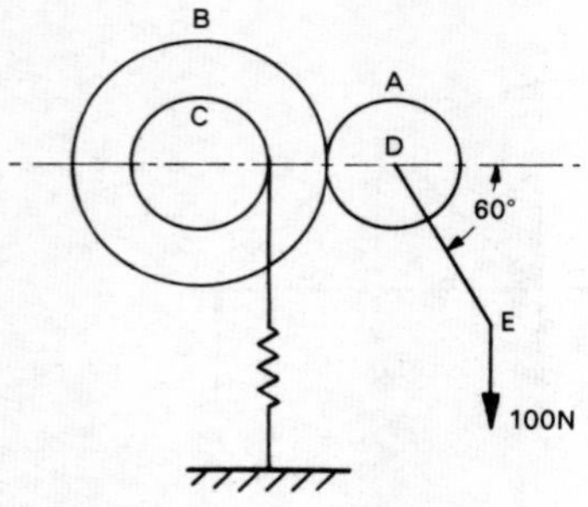

Ex. 4. 6

of 1 : 2. To wheel A is fixed an arm DE of length 0. 3 m and to wheel B is fixed a drum C of diameter 0. 2 m. A cord with a few turns round the drum is fixed to the drum at one end and is connected through a spring to an anchorage at the other end

When a vertical force of 100 N is applied to the end of the arm, the arm settles in equilibrium at 60° to the horizontal as shown. Calculate the tension in the spring, neglecting friction in the mechanism. [300 N]

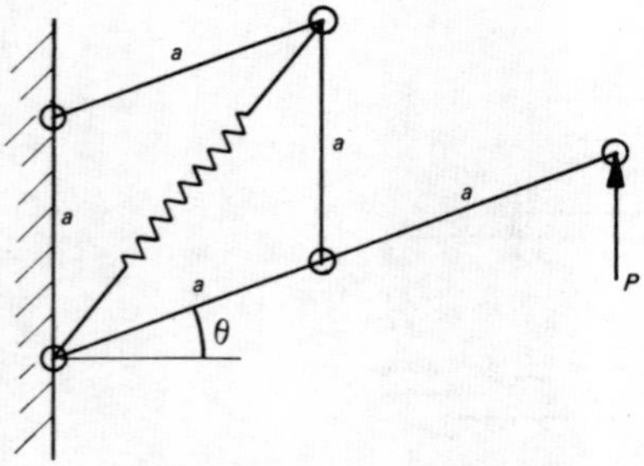

Ex. 4. 7

Ex. 4. 7 In the suspension shown the force in the spring (stiffness k) is zero when P and θ are zero.

Find the relationship between P and θ. $\left[P = \frac{k\,a}{2}\left(1 - (1 + \sin\,\theta)^{-1/2}\right)\right]$

Ex. 4.8 The structure shown has two effectively rigid members AB and BC of

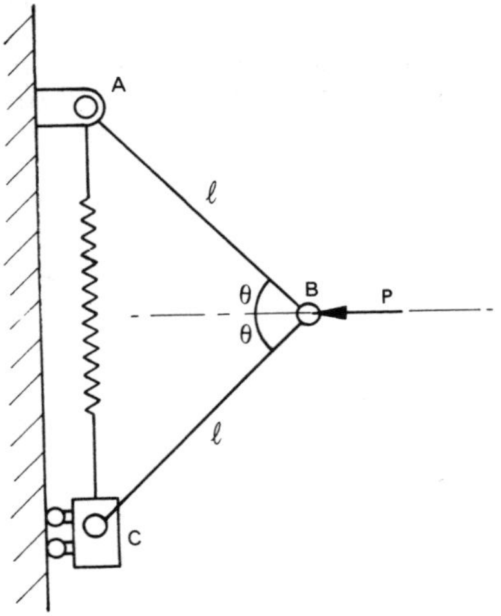

Ex. 4.8

equal length l. The third member is a spring of stiffness k. When there is no force in the spring the angle ABC is a right angle.

An increasing force P is applied as shown. Show that for equilibrium

$$P = 4lk\,(\sin\theta - \sin 45°)\cot\theta$$

Show also that under increasing P the structure becomes unstable when θ reaches about 63°. Neglect the weights of the members.

Ex. 4.9 A motor car jack consists of four similar links which are pin-jointed

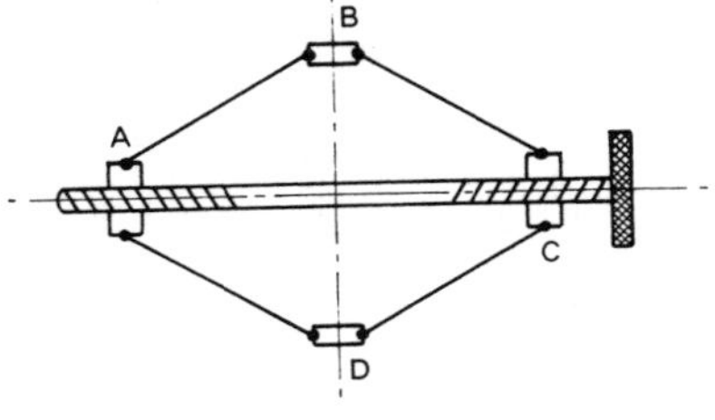

Ex. 4.9

at their ends to blocks A, B, C, D as shown. A shaft with single-start left-hand and right-hand screw threads connects opposite blocks A and C, so that turning the shaft draws these corners together, increasing the length of the vertical diagonal BD to lift the load.

What will be the tension in the shaft when the jack sustains a load of 1000 N with the links at 30° to the horizontal? Calculate the torque which must be applied to the shaft to lift the load in this position of the jack if the pitch of the threads is 2 mm. Neglect friction. [1732 N, 1.1 Nm]

5 Kinematics

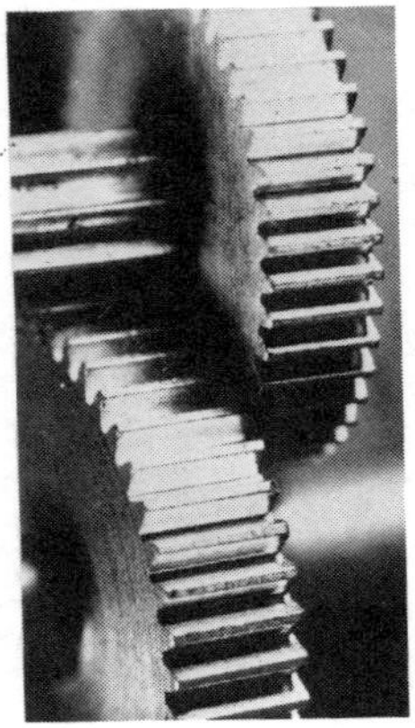

Kinematics is the study of motion, without reference to forces. To engineers it is the very important subject of the analysis (and much more difficult, synthesis) of mechanisms in terms of displacement, velocity and acceleration; as such it is the subject of many specialist texts.

Two-dimensional analysis and the supporting graphical constructions cover nearly all engineering kinematics, so this will be examined first, at some length. However, even in some quite simple mechanisms, the determination of accelerations requires the ability to describe motion in a moving reference frame. In vectors that is as easily treated in three dimensions as two, so relative motion will be used as an introduction to three-dimensional kinematics.

The vector nature of position, displacement, velocity and acceleration of a point has already been discussed in Chapter 2. In this chapter we shall consider the motion of bodies and assemblies of bodies.

Motion in a plane

The motion of a point in a fixed plane frame can be most easily visualised in rectangular coordinates Oxy with unit vectors $\mathbf{i}, \mathbf{j}$.

The position vector is $\mathbf{r}$, and the velocity and acceleration vectors are $\dot{\mathbf{r}}$ and $\ddot{\mathbf{r}}$, so that

$$\mathbf{r} = x\mathbf{i} + y\mathbf{j}$$
$$\dot{\mathbf{r}} = \dot{x}\mathbf{i} + \dot{y}\mathbf{j}$$
$$\ddot{\mathbf{r}} = \ddot{x}\mathbf{i} + \ddot{y}\mathbf{j}$$

In some situations it is more convenient to use polar coordinates r, θ. The time derivatives can be obtained by the use of unit vectors, say $\hat{\mathbf{r}}, \hat{\boldsymbol{\theta}}$, but although these have constant magnitude, they vary as vectors because their

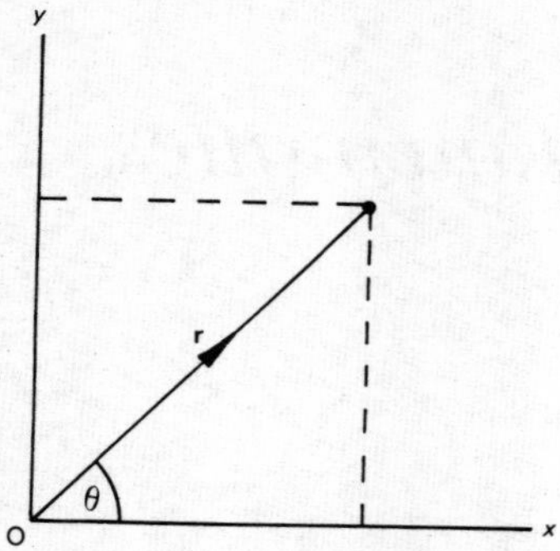

FIG. 5. 1

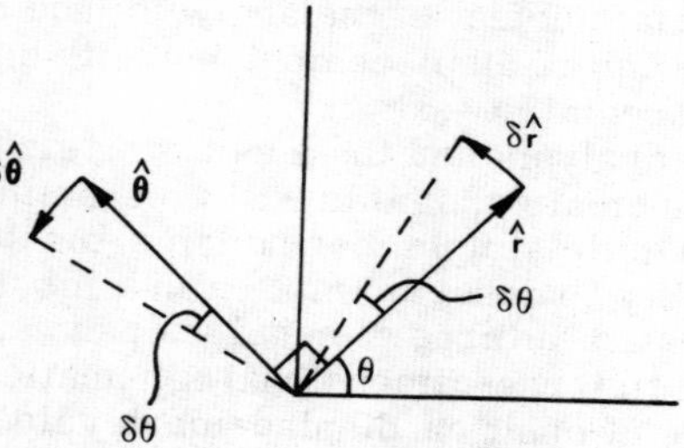

FIG. 5. 2 Rotating unit vectors

directions change with that of the position vector **r**. It can be seen from Fig. 5. 2 that

$$\delta\hat{\mathbf{r}} = \delta\theta\hat{\boldsymbol{\theta}} \text{ and } \delta\hat{\boldsymbol{\theta}} = -\delta\theta\hat{\mathbf{r}}$$

(jog your memory by referring to Fig. 2. 21)

so $$\dot{\hat{\mathbf{r}}} = \underset{\delta t \to 0}{\text{Lt}} \frac{\delta\hat{\mathbf{r}}}{\delta t} = \dot{\theta}\hat{\boldsymbol{\theta}} \text{ and } \dot{\hat{\boldsymbol{\theta}}} = -\dot{\theta}\hat{\mathbf{r}}$$

The position vector **r** can be written $\mathbf{r} = r\hat{\mathbf{r}}$.
Differentiating,

$$\begin{aligned}\dot{\mathbf{r}} &= \dot{r}\hat{\mathbf{r}} + r\dot{\hat{\mathbf{r}}} \\ &= \dot{r}\hat{\mathbf{r}} + r\dot{\theta}\hat{\boldsymbol{\theta}}\end{aligned} \qquad (5.1)$$

The radial and transverse components of the velocity are therefor $\dot{r}$ and $r\dot{\theta}$. Differentiating again,

$$\begin{aligned}\ddot{r} &= \ddot{r}\hat{\mathbf{r}} + \dot{r}\dot{\hat{\mathbf{r}}} + \dot{r}\dot{\theta}\hat{\boldsymbol{\theta}} + r\ddot{\theta}\hat{\boldsymbol{\theta}} + r\dot{\theta}\dot{\hat{\boldsymbol{\theta}}}\\ &= \ddot{r}\hat{\mathbf{r}} + \dot{r}\dot{\theta}\hat{\boldsymbol{\theta}} + \dot{r}\dot{\theta}\hat{\boldsymbol{\theta}} + r\ddot{\theta}\hat{\boldsymbol{\theta}} - r\dot{\theta}^2\hat{\mathbf{r}}\\ &= (\ddot{r} - r\dot{\theta}^2)\hat{\mathbf{r}} + (r\ddot{\theta} + 2\dot{r}\dot{\theta})\hat{\theta}\end{aligned}$$

So the radial and transverse components of the acceleration are $(\ddot{r} - r\dot{\theta}^2)$ and $r\ddot{\theta} + 2\dot{r}\dot{\theta}$.

Relative velocity and acceleration

If the motion of point P in a plane is described *relative* to the moving point Q by $\mathbf{r}$, then

$$\begin{aligned}\rho &= \mathbf{R} + \mathbf{r}\\ \dot{\rho} &= \dot{\mathbf{R}} + \dot{\mathbf{r}}\\ \ddot{\rho} &= \ddot{\mathbf{R}} + \ddot{\mathbf{r}}\end{aligned}$$

where $\mathbf{R}$ describes the motion of Q.

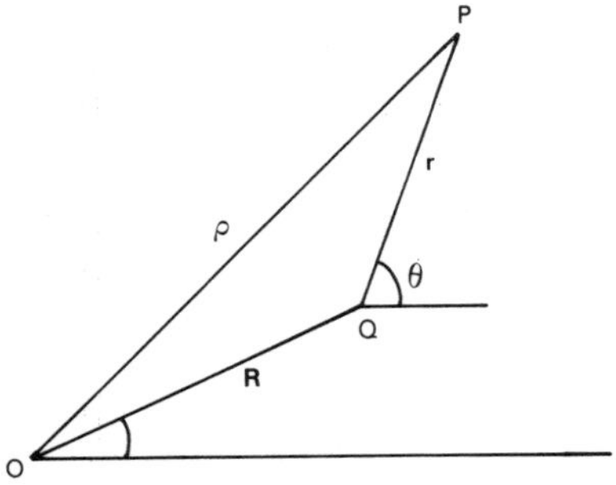

FIG. 5. 3

In words,

$$\text{the} \begin{pmatrix}\text{velocity}\\ \text{acceleration}\end{pmatrix} \text{of P} = \begin{pmatrix}\text{velocity}\\ \text{acceleration}\end{pmatrix} \text{of Q} + \begin{pmatrix}\text{velocity}\\ \text{acceleration}\end{pmatrix} \text{of P relative to Q}$$

This concept of relative velocity and acceleration is very useful in analysing the motion of a mechanism consisting of effectively rigid links. If the vector QP is used to describe a rigid link, the *length* remains constant, and so $\dot{r} = 0$. The velocity and acceleration of P relative to Q are therefore, from (5.1) and (5.2),

$$\begin{aligned}{}_{Q}\mathbf{v}_{P} &= {}_{Q}\mathring{\mathbf{r}}_{P} = r\dot{\theta}\hat{\boldsymbol{\theta}}\\ {}_{Q}\mathbf{f}_{P} &= {}_{Q}\ddot{\mathbf{r}}_{P} = -r\dot{\theta}^2\hat{\mathbf{r}} + r\ddot{\theta}\hat{\theta}\end{aligned}$$

where the suffix and prefix denote 'of P relative to Q'. These expressions confirm what is intuitively obvious—that one point on a rigid link can have only a tangential velocity relative to another point on the link, but will in general have a relative acceleration with two components, tangential $r\ddot{\theta}$ and radial $-r\dot{\theta}^2$; the latter, which is always radially *inwards*, is often called the centripetal component.

These statements can be shown very effectively on simple vector diagrams. Consider the motion in a plane of a single link QPC (Fig. 5.4 (a)). The velocity and acceleration diagrams are shown in Figs. 5.4 (b) and 5.4 (c).

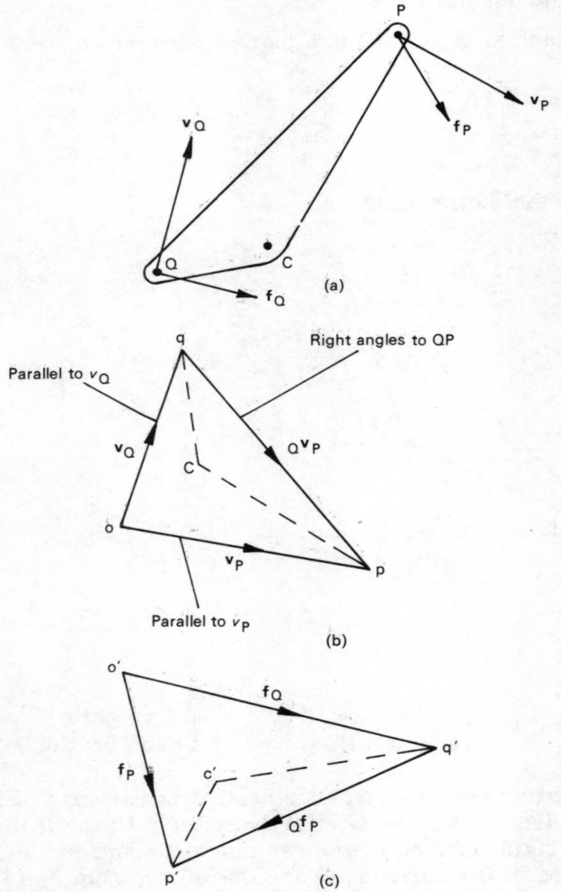

FIG. 5.4

(a) Link QPC (b) Velocity diagram (c) acceleration diagram

The construction of diagrams like these will be described in detail in the examples that follow, but for the present it should be noted that (a) the poles (o, o′) of the diagrams represent all points at rest on the velocity diagram, and all points with zero acceleration on the acceleration diagram, (b) vectors emanating from the poles represent *absolute* velocities or accelerations, whereas (c) other vectors represent *relative* velocities and accelerations. Diagrams drawn for complete mechanisms are often useful as aids to finding analytical expressions for velocity and acceleration; and for complicated mechanisms, accurately drawn diagrams provide a simple and sufficiently accurate escape from much complicated algebra.

The points c and c′ on the diagrams, representing the velocity and acceleration of the third point C on the link QPC can be found by drawing triangles qpc and q′p′c′ similar to triangle QPC, where cp is perpendicular to CP and c′p′ is parallel to CP because C must be moving in a circle relative to P. The validity of these simple constructions can be confirmed by the reader.

Velocities and accelerations of simple mechanisms

The engine mechanism is a good starting point because it can be easily treated analytically or graphically. It is shown diagrammatically in Fig. 5. 5.

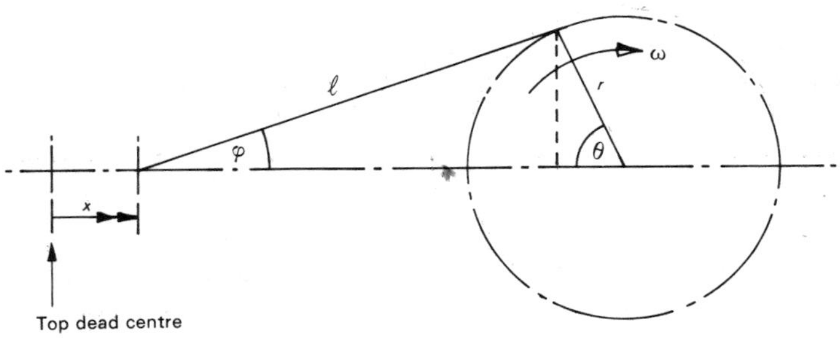

FIG. 5. 5 The engine mechanism

$$x = r + l - (r\cos\theta + l\cos\phi)$$

ϕ can be eliminated by noting that

$$r\sin\theta = l\sin\phi$$

So $$\sin\phi = \frac{r}{l}\sin\theta = \frac{\sin\theta}{n}$$

using n to denote $\frac{l}{r}$.

$$\cos\phi = (1-\sin^2\phi)^{1/2} = \left(1 - \frac{\sin^2\theta}{n^2}\right)^{1/2}$$

Hence

$$x = r + l - r[\cos\theta + (n^2 - \sin^2\theta)^{1/2}]$$

The velocity and acceleration of the piston can be found by differentiating

$$\dot{x} = \frac{dx}{d\theta}\cdot\frac{d\theta}{dt} = \omega\frac{dx}{d\theta}$$

$$\dot{x} = r\omega\left[\sin\theta + \frac{\sin^2\theta}{2(n^2-\sin^2\theta)^{1/2}}\right] \tag{5.3}$$

The expression for $\ddot{x}$ is very cumbersome, but is simplified for the case of ω = constant. For constant ω

$$\ddot{x} = \frac{d\dot{x}}{d\theta}\frac{d\theta}{dt}$$

$$\ddot{x} = r\omega^2\left[\cos\theta + \frac{n^2\cos 2\theta + \sin^4\theta}{2(n^2-\sin^2\theta)^{3/2}}\right]$$

These expressions are usually simplified for further analysis (of small amplitude torsional vibrations for example) by expanding them as binomials, leading to

$$\dot{x} = r\omega\left[\sin\theta + \frac{1}{2n}\sin 2\theta + \frac{1}{8n^3}(\sin 2\theta - \frac{1}{2}\sin 4\theta) + \text{etc.}\right]$$

$$\ddot{x} = r\omega^2\left[\cos\theta + \frac{1}{n}\cos 2\theta + \frac{1}{4n^3}(\cos 2\theta - \cos 4\theta) + \text{etc.}\right]$$

The value of n in internal combustion engines is usually in the region of 3–4, so for many purposes the first two terms in each of the expressions will suffice.

These analytical results for the very simple crank mechanism of the reciprocating engine show the sort of complexity that is in store in more elaborate mechanisms, and arouse an interest in the graphical method. For example, finding the acceleration of the centre of gravity of the connecting rod in the engine mechanism is very much easier graphically than analytically. However, it must be remembered that an analytical expression applies to all positions of the mechanism while a velocity or acceleration diagram applies only to one position, so it may be necessary to construct a number of vector diagrams.

The vector diagrams are easily drawn for the crank mechanism. If we assume that the crank angular velocity ω is constant and known, we proceed as follows:

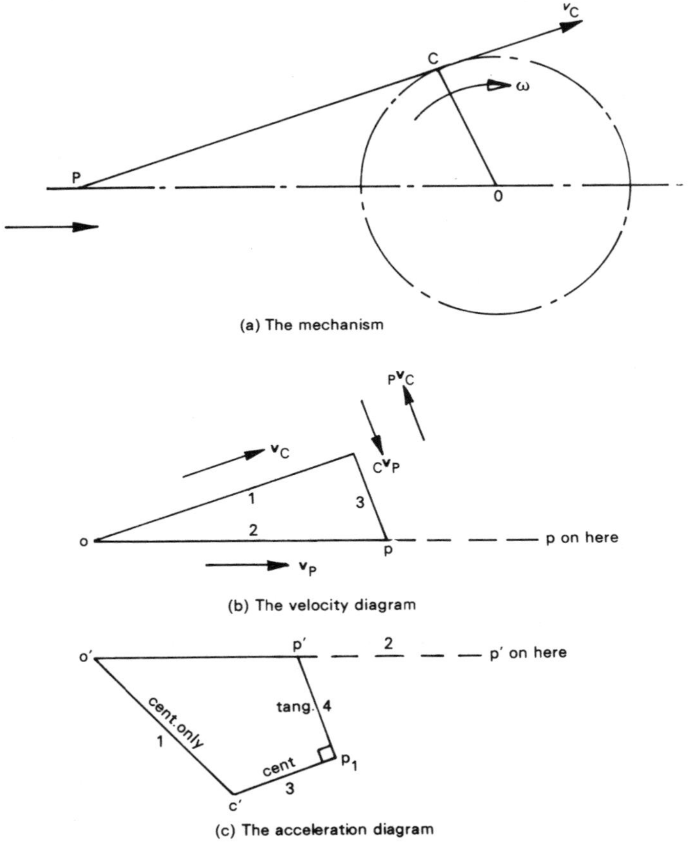

(a) The mechanism

(b) The velocity diagram

(c) The acceleration diagram

FIG. 5. 6

Figure 5. 6 (a) shows the position of the mechanism chosen for analysis. With ω and the dimensions known, $\mathbf{v}_C$, the velocity of the crank pin, can be calculated. For the chosen position, the line oc can be drawn in Fig. 5. 6 (b) to represent the vector $\mathbf{v}_C$. Next the *direction* of the line op can be set down because P is constrained to move along the line PO. Finally the velocity diagram can be completed by the line cp which represents the velocity of P *relative to* C and which must therefore be perpendicular to PC. The intersection locates p.

Figure 5. 6 (c) is the acceleraticn diagram constructed in much the same way. Because ω is constant, C has no tangential acceleration, but it does have a centripetal acceleration $OC\omega^2 (= oc^2/OC)$ which is shown as o′c′. P can only move along the straight line PO so it can only have an acceleration in that direction

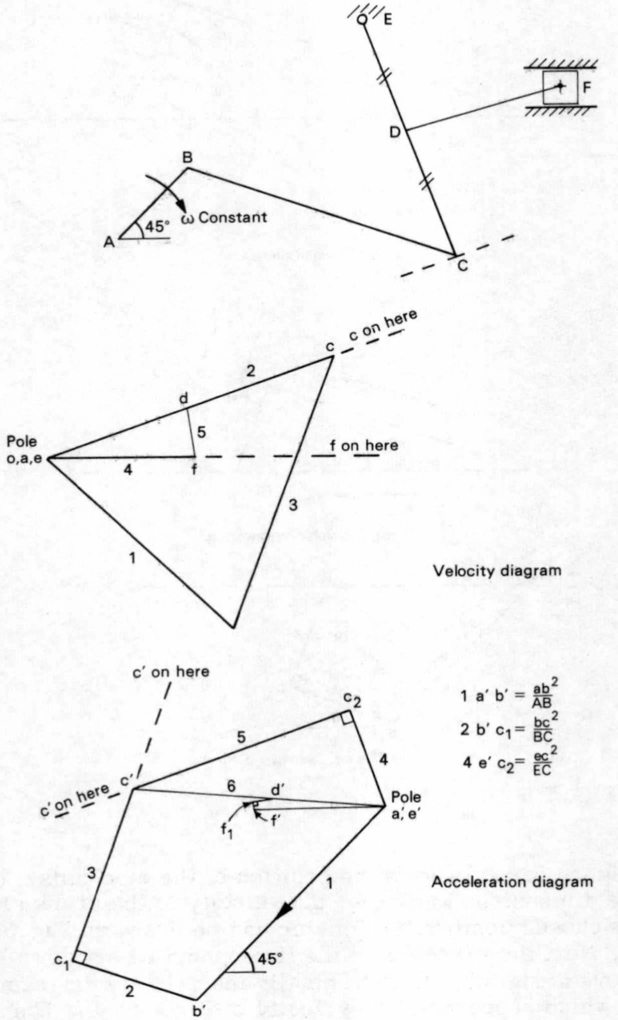

FIG. 5. 7

so a line along which p′ must be located can next be drawn. The diagram is closed by the acceleration of P relative to C, which will appear as c′p′ on the diagram; this acceleration will in general have two components, centripetal and tangential. Now the first of these is $\frac{cp^2}{CP}$ (the $-r\dot{\theta}^2$ term) in magnitude and its direction is the direction PC, and it is drawn next as $c'p_1$; the second is the $r\ddot{\theta}$ term and its magnitude cannot be found from the velocity diagram, but its direction is perpendicular to PC and therefore to $c'p_1$ and its intersection with o′p′ gives its magnitude and locates p′ on the acceleration diagram. The acceleration of P is given by o′p′ in magnitude and direction.

The scales of these diagrams must be chosen to suit the accuracy required in the answer, but with care and appropriate drawing instruments high accuracy can be achieved. A final reminder is that *absolute* velocities and accelerations are given by vectors emanating from the respective origins o and o′.

Figure 5.7 gives another example of velocity and acceleration diagrams. The numbers on the lines show the order in which they were drawn. In each case a start is made from a point in the mechanism whose velocity and acceleration are known.

Instantaneous centres

The velocities and accelerations of mechanisms are examined in this chapter chiefly in terms of vector diagrams of the sort described in the last section. An alternative to the velocity diagram, which will only be mentioned here, is the use of instantaneous centres. The method of instantaneous centres can be most useful in simple cases but can become very cumbersome in complex mechanisms.

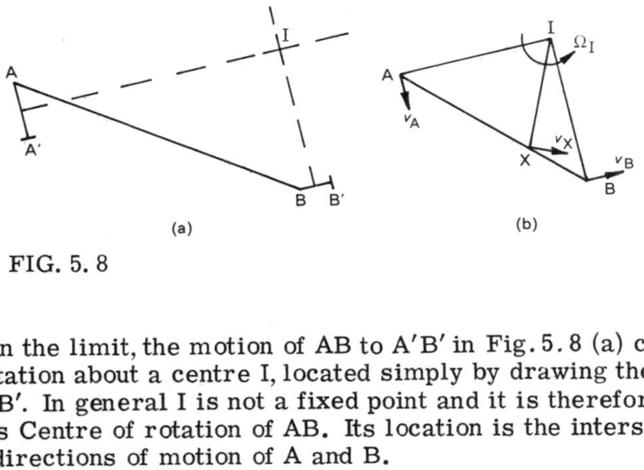

FIG. 5. 8

In the limit, the motion of AB to A′B′ in Fig. 5. 8 (a) can be regarded as a pure rotation about a centre I, located simply by drawing the bisecting normals to AA′ and BB′. In general I is not a fixed point and it is therefore called the Instantaneous Centre of rotation of AB. Its location is the intersection of the normals to the directions of motion of A and B.

It is plain from Fig. 5. 8 (b) that the instantaneous angular velocity about I is given by

$$\Omega_I = v_A/IA$$

$$\Omega_I = v_B/IB$$

and that $$v_B = v_A \frac{IB}{IA}$$

The velocity of any other point on the body, for example X, can then be found,

$$v_X = \Omega_I \, IX = v_A \frac{IX}{IA}$$

and its direction is perpendicular to IX.

Some simple, self-evident examples are shown in Fig. 5. 9; a four-bar chain, a rolling wheel and the slipping ladder, loved by school-teachers if not by window-cleaners.

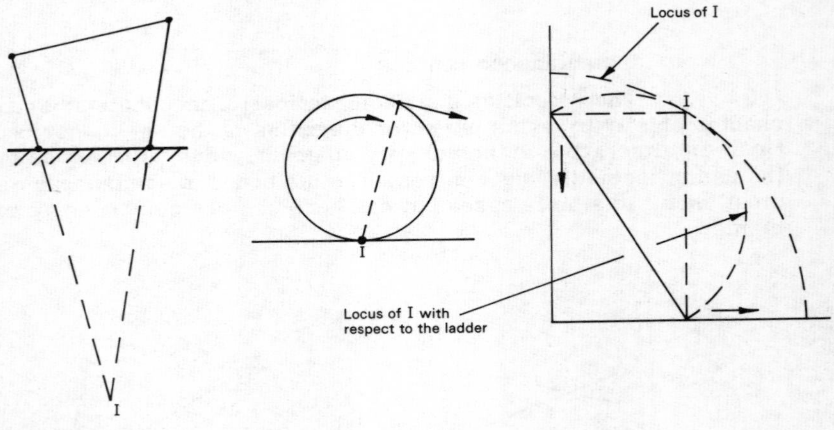

FIG. 5. 9

The dotted lines on the ladder example are the loci of the instantaneous centre; the larger radius quarter-circle is the locus with respect to fixed axes (the wall and the floor), and the smaller radius semi-circle is the locus drawn with respect to the moving ladder. These loci are called respectively 'space and body centrodes' and it is evident from the diagram that the body centrode rolls on the space centrode as the ladder moves. This and other very interesting properties of centrodes are exploited in specialised texts on kinematics.

One of the prettiest simple examples of instantaneous centre is the engine mechanism.

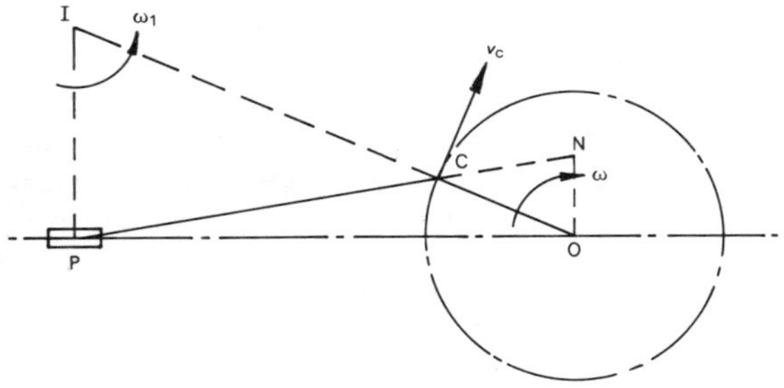

FIG. 5. 10

The instantaneous centre of the connecting rod PC in Fig. 5.10 is plainly at the intersection of the normal to the piston motion and the extension of the crank. If the angular velocity of the crank is ω and the instantaneous angular velocity of the connecting rod is ω_1, then the crank and piston velocities are related as follows:

$$v_c = \mathrm{OC}\,\omega = \mathrm{IC}\,\omega_1$$
$$v_p = \mathrm{IP}\,\omega_1 = \frac{\mathrm{IP}}{\mathrm{IC}}\,\mathrm{OC}\,\omega$$

If we now produce PC to meet the vertical through O at N, the triangles IPC and ONC are similar and

$$\frac{\mathrm{IP}}{\mathrm{IC}}\mathrm{OC} = \mathrm{ON}$$

So

$$v_p = \mathrm{ON}\,\omega$$

The piston velocity v_p is approximately (very nearly) a maximum when the angle PCO is a right angle. If the triangle OCN in Fig. 5.10 is compared with the velocity diagram in Fig. 5.6 (b) it can be seen that they are similar triangles, OCN being rotated through a right angle relative to OCP. In other words, OCN is a velocity diagram for the mechanism; similarly, of course, so is triangle ICP.

It will be recalled from the previous chapter that, in the absence of inertia effects, the torque on the crank of the engine mechanism is related to the force on the piston by

$$T = F\frac{v_p}{\omega}$$

It is therefore given by

$$T = F\,\mathrm{ON}$$

where ON is the length shown in Fig. 5.10. So the diagram can be used for estimating torque as well as velocity.

Gear trains

A simple but important special case of the use of relative velocities is the analysis of gear trains, where interest is almost entirely in angular velocities. The geometry of toothed gear wheels will not be examined in this book, but it is sufficient for our purpose to say that a mating pair of toothed wheels, as in Fig. 5.11, can be represented kinematically by a pair of pitch circles which roll on each other without slip; the numbers of teeth on the mating wheels must be proportional to the pitch circle radii and of course there must be a whole number of teeth on any wheel. The angular velocities are related by the connection that the

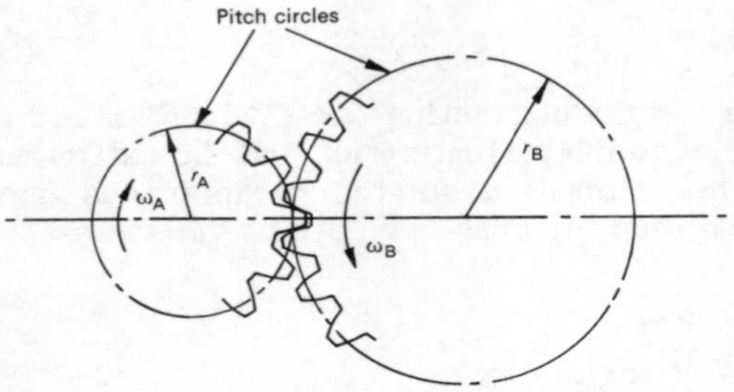

FIG. 5.11

pitch circles have the same linear velocity. If we use the convention that clockwise angular velocity is positive, then

$$r_A \omega_A = -r_B \omega_B$$

$$\omega_A / \omega_B = -r_B / r_A$$

or

$$\frac{\omega_A}{\omega_B} = -\frac{B}{A}, \tag{5.4}$$

where A and B are the numbers of teeth on wheels A and B, because for the teeth to mesh $\frac{r_A}{A} = \frac{r_B}{B}$.

Gear trains can be broadly classified as *ordinary*, when all the wheel centres are fixed, or *epicyclic*, when one or more of the wheel centres themselves rotate about the axis of another wheel. Within both these divisions there are *simple* gear trains, where each shaft carries only one wheel, and *compound* trains, where at least one of the wheels is a 'compound' wheel consisting essentially of two wheels with different numbers of teeth mounted on the same shaft and running at the same angular velocity. Diagrammatic examples are shown in Fig. 5.12.

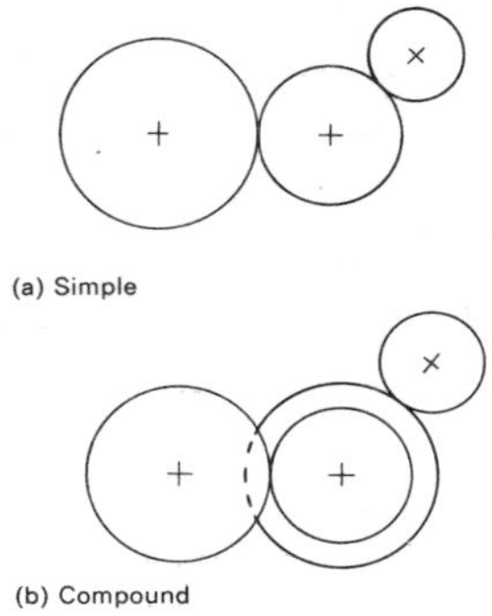

FIG. 5.12 Ordinary gear trains

Calculating the speeds and directions of the various wheels, in terms of numbers of teeth or pitch circle radii, is very simple for ordinary gear trains. For epicyclic trains, the calculation is an interesting example of relative velocities.

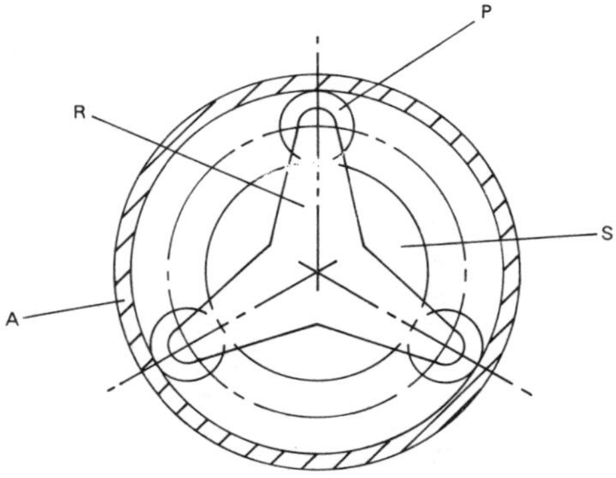

FIG. 5.13 A simple epicyclic gear train

In the train illustrated in Fig. 5.13 there are four kinematic elements, (a) the annulus, A, an internally toothed wheel, (b) the sun wheel, S, (c) the three planets, P, and (d) the mounting for the planets, R, variously described as the arm, the spider or simply the planet carrier.

The annulus, the sun and the arm are coaxial, but mounted on different shafts, and of course the planets' axes rotate about the common axis of the other three elements. Various input/output motions can be achieved by driving and/or constraining two of the three coaxial elements, and taking the output from the third. It is common to fix one of the elements: for example a band brake applied to the annulus would allow the arm and the sun to be used as driver and driven shafts; this arrangement achieves the 'clutching' action and a speed ratio, and a number of these side-by-side can be used as the basis of a vehicle gear box. An example is shown in Fig. 5.14. In this complicated arrangement, four forward speed ratios and one reverse can be selected by braking the various outer rings: 4.2 to 1, 2.4 to 1, 1.6 to 1 reduction and 1 to 1 (straight through drive); and 6.1 to 1 reverse.

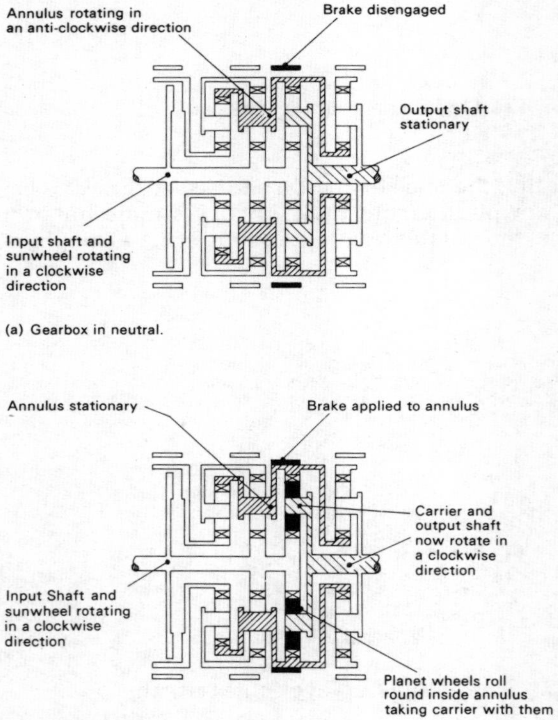

(a) Gearbox in neutral.

(b) Drive engaged (1st speed)

*FIG. 5.14

The various speed ratios in a single epicyclic train like that of Fig. 5.13 can easily be calculated if all the angular velocities are measured relative to the arm. To do this imagine yourself sitting on the arm and going round with it; to you the arm (and therefore the planet axes) will appear stationary and the gear train will look like an *ordinary* gear train, with all the axes fixed. The annulus will be rotating around you in one direction and the sun in the opposite direction. Remembering that what you see, including the angular velocities, is relative to the angular velocity of the arm, the velocity ratios can be written down directly; in terms of the number of teeth on the wheels, and having due regard for sign (by calling all clockwise angular velocities positive, for instance) equation (5.4) gives

$$\frac{\omega_A - \omega_R}{\omega_P - \omega_R} = \frac{P}{A}$$

$$\frac{\omega_P - \omega_R}{\omega_S - \omega_R} = -\frac{S}{P}$$

where A, S and P denote numbers of teeth. The overall speed ratio is given by

$$\frac{\omega_A - \omega_R}{\omega_S - \omega_R} = -\frac{S}{A} \tag{5.5}$$

Common special cases are fixed annulus and fixed sun. With the annulus fixed the speed ratio for arm and sun is obtained by putting $\omega_A = 0$ in (5.5)

$$\frac{0 - \omega_R}{\omega_S - \omega_R} = -\frac{S}{A}$$

giving $$\frac{\omega_S}{\omega_R} = 1 + \frac{A}{S}$$

with the sun fixed

$$\frac{\omega_A - \omega_R}{0 - \omega_R} = -\frac{S}{A}$$

$$\frac{\omega_A}{\omega_R} = 1 + \frac{S}{A}$$

Moving reference frames

> 'Hitherto I have laid down the definitions of such words as are less known, and explained the sense in which I would have them to be understood in the following discourse. I do not define time, space, place, and motion, as being well known to all. Only I must observe, that the common people conceive those quantities under no other notions but from the relation they bear to sensible objects. And thence arise, certain prejudices, for the removing of which it will be convenient to distinguish them into absolute and relative, true and apparent, mathematical and common.'
>
> *Principia*, Newton

In the last section and earlier in this chapter we have stationed observers on moving parts of a machine and then talked about 'the velocity and acceleration of this part relative to that part' and so on. Implicit in all this, somewhere in the background, was some 'fixed' frame of reference, Newton's 'absolute space', against which 'absolute' velocities and accelerations could be measured. In the prevailing models of the universe it is evident that no such absolute frame of reference exists. One must resort therefore to frames of reference which are 'fairly absolute' or 'absolute enough' in a particular situation. A bit of judgement is required. In analysing the motion of a toffee-wrapping machine or a family motor-car, the earth can be taken as an absolute reference, but in calculating the trajectory of long-range gunfire the earth's rotation must be taken into account; the calculation of the flight-path of an interplanetary rocket must employ a reference frame that accommodates the whole solar system at least. It is not always simply a matter of scale; the working of the gyrocompass is dependent on the rotation of the earth.

Even in some very simple earth-bound mechanisms it is convenient to express motions in terms of the sum of relative motions, as anyone who has been to a fairground and ridden on the 'Waltzer' will appreciate. We shall therefore embark next on an analysis of motion which is described within a moving reference frame. This is one of those cases where it is as easy to do it in three dimensions as in two, so we shall do it in three. But first it is worth looking at a few simple cases in two dimensions to get the feel of it. In the general treatment the kinematics gets lost for some of the time in the algebra.

Life on a roundabout

Imagine that an unusually tolerant fairground proprietor allows you to experiment with one of his simple roundabouts; it is a flat turntable which rotates in a horizontal plane. There are two positions from which you can make observations; one is standing still in the hole in the middle, with the roundabout rotating around you, the other is standing on the roundabout, going round with it.

Suppose now you set up a clockwork train set, first with a straight track mounted along a radius. Consider the motion of the toy engine running

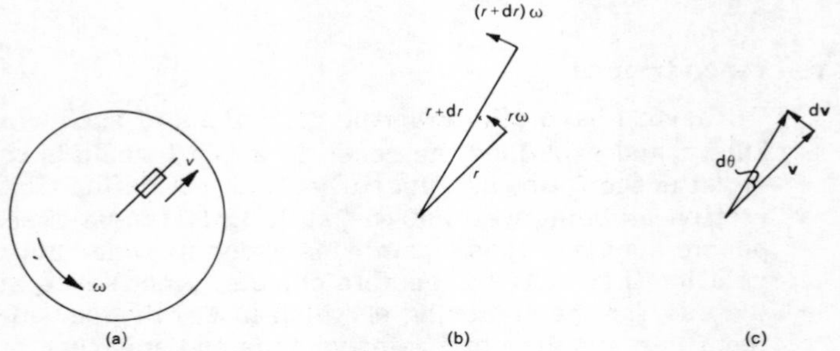

FIG. 5.15

outwards along the radius at constant speed v, while the roundabout rotates at constant angular velocity ω. The essentials are shown in Fig. 5.15. If you stand on the roundabout the engine will appear to be moving in a straight line at constant velocity, in other words with no acceleration. If you then stand in the hole in the middle, you become a stationary observer and see the engine moving along a spiral path. This is a complicated looking motion and one now sees the possible simplifications of describing the motion in such terms as 'the velocity or acceleration of that plus the velocity or acceleration of this relative to that'. We must be careful however, because we have no foundation for assuming that we can add two vectors in different frames of reference. Our conclusions earlier in the chapter about relative velocities and accelerations were based on the description of the motion of points in the same fixed frame.

Any point on the roundabout, including the point at which the engine is situated at any instant, has only one component of acceleration, the centripetal component $-r\omega^2$. *Relative to the roundabout* the engine has no acceleration. Does this mean therefore, that the engine has only the centripetal acceleration $-r\omega^2$? Obviously it does not—a moment's thought about the tangential velocity reveals that it is increasing, as Fig. 5.15 (b) shows, simply because the radius is increasing, and an increasing tangential velocity means a tangential acceleration; the tangential velocity is evidently increasing at a rate $\omega \mathrm{d}r/\mathrm{d}t = \omega v$. A few more moments' (somewhat deeper) thought reveals another component of tangential acceleration with the same magnitude; this arises from the fact that the velocity $\mathbf{v}$ of the engine which appears constant on the roundabout is actually changing direction with respect to the fixed observer. Figure 5.15 (c) shows that the *magnitude* of $\mathbf{dv}$ is $v\,\mathrm{d}\theta = v\omega\,\mathrm{d}t$, and so $\dfrac{\mathrm{d}\mathbf{v}}{\mathrm{d}t} = v\omega$ and its direction is tangential. So we have a total tangential component of acceleration equal to $2\omega v$. For this arrangement of the train set this component of acceleration is the $2\dot{r}\dot{\theta}$ of equation (5.2).

The tangential acceleration of the engine $2\omega v$ must be engendered by a tangential force between the track and the engine. If this force were not provided (if for instance there was no track and the engine was simply set going on a radial line on the smooth surface of the roundabout) the engine would drift away from the radial line in the direction opposite to the rotation. To the observer *on* the roundabout the engine would *seem* to be experiencing a tangential force pushing it sideways away from the radial line. Of course there is no force in the tangential direction—it is the *absence* of a tangential force which causes the drift.

The shell fired from a gun experiences this same drift effect due to the rotation of the earth. The size of the effect depends on where the gun is on the earth and the direction in which it is pointing; it is always small, but it is big enough to be worth taking into account in the trajectory calculations for long-range gunfire. Professional artillery men still describe the effect as 'the Coriolis force', although they know very well that the effect is not due to a force. The name Coriolis is that of the man who first drew attention to this component of acceleration in a motion, described with reference to a moving frame, Gaspard Gustave de Coriolis (1792-1843), a Frenchman variously described as mathematician and military engineer.

Returning now to the roundabout, suppose next that you set up the train set with a circular track, with its centre at the centre of the roundabout, as in Fig. 5.16. There are no complications here; the engine moves in a circle, its absolute

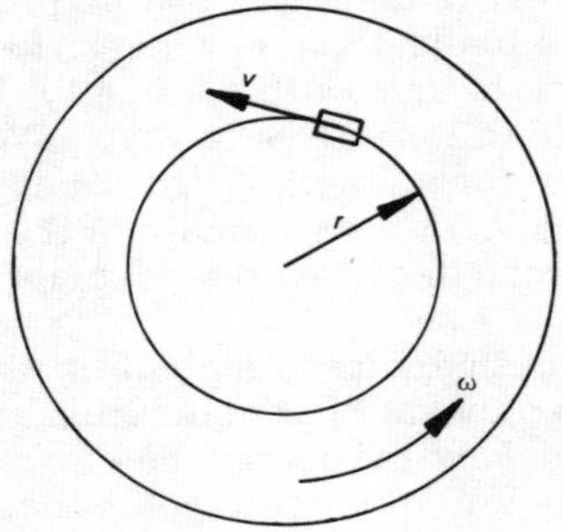

FIG. 5.16

velocity is $(r\omega + v)$ and its velocity relative to the roundabout, as seen by an observer on the roundabout, is v, in the same circle. Plainly its acceleration is purely centripetal and is

$$\frac{(r\omega + v)^2}{r} = r\omega^2 + 2v\omega + \frac{v^2}{r}$$

The first term on the right is the acceleration of the track, the last is the centripetal acceleration of the engine as measured by the observer on the roundabout, in other words the acceleration relative to the roundabout. The second term is $2v\omega$ again—Coriolis. We see that in both these simple cases the total acceleration can be described as: the acceleration of the point on the roundabout where the engine is located at the instant considered, plus the acceleration of the engine relative to the roundabout, plus $2v\omega$. Also in both cases the direction of the $2v\omega$ term is perpendicular to the direction of v, the relative velocity. We shall see below that the general result corresponding to this is just the same; the Coriolis component of acceleration is perpendicular to both the velocity relative to the moving frame and the $\boldsymbol{\omega}$ vector of the moving frame itself.

General relative motion

After that preamble, let us now examine any general motion of a point within a three-dimensional frame which is itself moving, its movement being defined in an absolute frame. This is plainly a case for vector treatment. Let the absolute frame be described by fixed axes XYZ with unit vectors $\mathbf{i}, \mathbf{j}, \mathbf{k}$ and the moving frame by axes xyz with unit vectors $\hat{\mathbf{i}}, \hat{\mathbf{j}}, \hat{\mathbf{k}}$. $\mathbf{i}, \mathbf{j}, \mathbf{k}$ are true constants but $\hat{\mathbf{i}}, \hat{\mathbf{j}}, \hat{\mathbf{k}}$, although of course constant in magnitude, vary in direction and therefore will have time derivatives. Figure 5.17 shows the axes xyz with origin at position $\mathbf{R}$ in

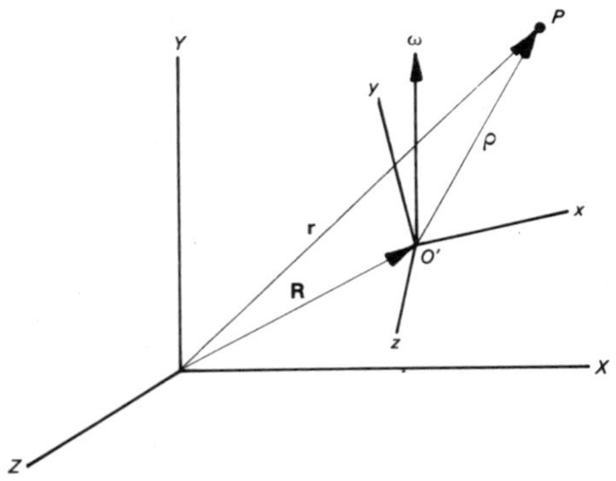

FIG. 5. 17

the *XYZ* frame, and rotating with angular velocity $\boldsymbol{\omega}$. The plan is to describe the motion of a point P in two ways: first as seen by an observer in the fixed frame, and second as seen by an observer in the moving frame, but corrected by the necessary additional terms to allow for the movement of the frame—as we did for the two simple cases on the roundabout. This may seem to be a rather elaborate exercise at the moment, but we shall see later that this is the only way to analyse the motion of many quite simple mechanisms.

First the motion of P can be specified by the vector **r** and its derivatives. These can be expressed in terms of the unit vectors **i, j, k** which are constants,

$$\mathbf{r} = X\mathbf{i} + Y\mathbf{j} + Z\mathbf{k}$$
$$\dot{\mathbf{r}} = \dot{X}\mathbf{i} + \dot{Y}\mathbf{j} + \dot{Z}\mathbf{k}$$
$$\ddot{\mathbf{r}} = \ddot{X}\mathbf{i} + \ddot{Y}\mathbf{j} + \ddot{Z}\mathbf{k}$$

For the second description of the absolute motion of P we write

$$\mathbf{r} = \mathbf{R} + \boldsymbol{\rho}$$

and differentiate with respect to time to get the velocity and acceleration. Note that in doing this we are treating all three as vectors in the *XYZ* frame.

$$\dot{\mathbf{r}} = \dot{\mathbf{R}} + \dot{\boldsymbol{\rho}}$$

$\dot{\mathbf{R}}$ is the velocity of the origin of the *xyz* coordinates (the moving frame).
To find out what $\dot{\boldsymbol{\rho}}$ means we can define the vector $\boldsymbol{\rho}$ in terms of unit vectors in the *xyz* frame:

$$\boldsymbol{\rho} = x\hat{\mathbf{i}} + y\hat{\mathbf{j}} + z\hat{\mathbf{k}}$$

Remembering that $\hat{\mathbf{i}}, \hat{\mathbf{j}}, \hat{\mathbf{k}}$ vary in direction,

$$\dot{\boldsymbol{\rho}} = (\dot{x}\hat{\mathbf{i}} + \dot{y}\hat{\mathbf{j}} + \dot{z}\hat{\mathbf{k}}) + (x\dot{\hat{\mathbf{i}}} + y\dot{\hat{\mathbf{j}}} + z\dot{\hat{\mathbf{k}}})$$

Now $\mathring{\hat{\mathbf{i}}} = \boldsymbol{\omega} \wedge \hat{\mathbf{i}}, \mathring{\hat{\mathbf{j}}} = \boldsymbol{\omega} \wedge \hat{\mathbf{j}}, \mathring{\hat{\mathbf{k}}} = \boldsymbol{\omega} \wedge \hat{\mathbf{k}}$

so $x\mathring{\hat{\mathbf{i}}} + y\mathring{\hat{\mathbf{j}}} + z\mathring{\hat{\mathbf{k}}} = \boldsymbol{\omega} \wedge (x\hat{\mathbf{i}} + y\hat{\mathbf{j}} + z\hat{\mathbf{k}}) = \boldsymbol{\omega} \wedge \boldsymbol{\rho}$

Also $(\dot{x}\hat{\mathbf{i}} + \dot{y}\hat{\mathbf{j}} + \dot{z}\hat{\mathbf{k}})$ is the velocity of P as it would be seen by an observer on the *xyz* frame, i.e. the velocity of P *relative* to the *xyz* frame, which we will call $\dot{\boldsymbol{\rho}}_r$. So we see that

$$\mathring{\boldsymbol{\rho}} = \boldsymbol{\omega} \wedge \boldsymbol{\rho} + \dot{\boldsymbol{\rho}}_r$$

and
$$\dot{\mathbf{r}} = \underbrace{\dot{\mathbf{R}} + \boldsymbol{\omega} \wedge \boldsymbol{\rho}} + \dot{\boldsymbol{\rho}}_r \qquad (5.6)$$

The velocity of a point, Q say, fixed to the moving frame and coincident with P at the instant considered. — The velocity of P relative to Q.

Differentiating again with $\dot{\boldsymbol{\rho}}_r$ in the form $(\dot{x}\hat{\mathbf{i}} + \dot{y}\hat{\mathbf{j}} + \dot{z}\hat{\mathbf{k}})$ leads to

$$\ddot{\mathbf{r}} = \ddot{\mathbf{R}} + \dot{\boldsymbol{\omega}} \wedge \boldsymbol{\rho} + \boldsymbol{\omega} \wedge \mathring{\boldsymbol{\rho}} + (\ddot{x}\hat{\mathbf{i}} + \ddot{y}\hat{\mathbf{j}} + \ddot{z}\hat{\mathbf{k}}) + (\dot{x}\mathring{\hat{\mathbf{i}}} + \dot{y}\mathring{\hat{\mathbf{j}}} + \dot{z}\mathring{\hat{\mathbf{k}}}) \qquad (5.7)$$

The third term can be rewritten

$$\boldsymbol{\omega} \wedge \mathring{\boldsymbol{\rho}} = \boldsymbol{\omega} \wedge (\boldsymbol{\omega} \wedge \boldsymbol{\rho}) + \boldsymbol{\omega} \wedge \dot{\boldsymbol{\rho}}_r$$

The fourth term is the acceleration of P as it would be seen by the observer on the *xyz* frame, so it can appropriately be called $\ddot{\boldsymbol{\rho}}_r$, the acceleration of P relative to the *xyz* frame.

The last term can be simplified

$$(\dot{x}\mathring{\hat{\mathbf{i}}} + \dot{y}\mathring{\hat{\mathbf{j}}} + \dot{z}\mathring{\hat{\mathbf{k}}}) = \boldsymbol{\omega} \wedge (\dot{x}\mathbf{i} + \dot{y}\hat{\mathbf{j}} + \dot{z}\hat{\mathbf{k}}) = \boldsymbol{\omega} \wedge \dot{\boldsymbol{\rho}}_r$$

The whole thing, (5. 7), can now be written

$$\ddot{\mathbf{r}} = \ddot{\mathbf{R}} + \boldsymbol{\omega} \wedge (\boldsymbol{\omega} \wedge \boldsymbol{\rho}) + \dot{\boldsymbol{\omega}} \wedge \boldsymbol{\rho} \quad \text{(The acceleration of Q)}$$
$$+ \ddot{\boldsymbol{\rho}}_r \quad \text{(The acceleration of P relative to Q)}$$
$$+ 2\boldsymbol{\omega} \wedge \dot{\boldsymbol{\rho}}_r \quad \text{(The Coriolis component of the acceleration of P)}$$
(5.8)

In equations (5. 6) and (5. 8) the point referred to as Q is a point *fixed* to the moving frame but coincident with P at the instant considered.

This is a cumbersome expression so perhaps it would be as well to look at a two-dimensional example before we tackle one in three. A common mechanical application is a slider moving along a curved or straight track which is itself rotating. Figure 5.18 shows a sketch of part of a mechanism consisting of a curved track lying in the plane of the paper and rotating about an axis through O perpendicular to the paper. A slider P slides along the track with some velocity and acceleration relative to the track. What will be the absolute velocity and acceleration of P? (This example could easily be set up with the train set and roundabout).

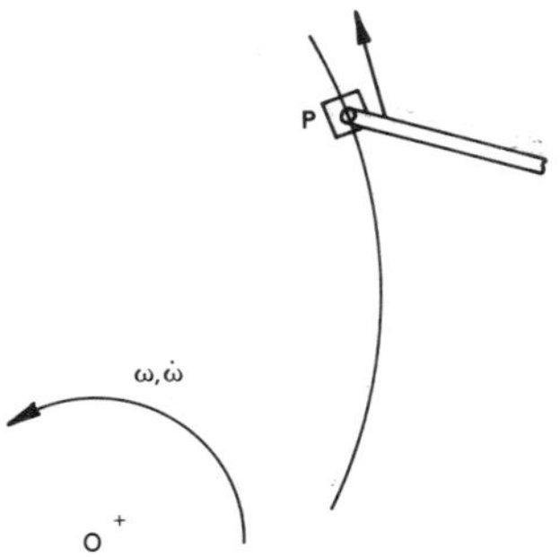

FIG. 5.18

The first thing to do is to reduce (5.8) to the simple case of the *xyz* axes rotating about a single fixed axis through O. The sensible thing to do is to take the *Z* and *z* axes as coincident and as the axis of rotation. **R** is zero (as are $\dot{\mathbf{R}}$ and $\ddot{\mathbf{R}}$) and $\boldsymbol{\omega}$ is simply the rotation of the *xy* axes about the *Z*/*Z* axis. If we recall the triple vector product (2.5)

$$\mathbf{A} \wedge (\mathbf{B} \wedge \mathbf{C}) = (\mathbf{A}.\mathbf{C})\mathbf{B} - (\mathbf{A}.\mathbf{B})\mathbf{C}$$

then

$$\begin{aligned} \omega \wedge (\omega \wedge \boldsymbol{\rho}) &= (\boldsymbol{\omega}.\boldsymbol{\rho})\boldsymbol{\omega} - (\boldsymbol{\omega}.\boldsymbol{\omega})\boldsymbol{\rho} \\ &= (\boldsymbol{\omega}.\boldsymbol{\rho})\boldsymbol{\omega} - \omega^2\boldsymbol{\rho} \end{aligned}$$

Because the *z* and *Z* axes are coincident (parallel would be enough), $\boldsymbol{\omega}$ and $\boldsymbol{\rho}$ are perpendicular, so $\boldsymbol{\omega}.\boldsymbol{\rho} = 0$ and

$$\omega \wedge (\boldsymbol{\omega} \wedge \boldsymbol{\rho}) = -\omega^2\boldsymbol{\rho} \quad \text{for the plane case.}$$

(5.8) therefore reduces to

$$\ddot{\mathbf{r}} = -\omega^2\boldsymbol{\rho} + \dot{\boldsymbol{\omega}} \wedge \boldsymbol{\rho} + \ddot{\boldsymbol{\rho}}_{\mathrm{r}} + 2\omega \wedge \dot{\boldsymbol{\rho}}_{\mathrm{r}}$$

centripetal transverse
components of acceleration of
Q, fixed to slideway and
instantaneously coincident with P.

The directions of the vector product terms are easy to see using the right-hand screw rule. Both are perpendicular to $\boldsymbol{\omega}$ (and $\dot{\boldsymbol{\omega}}$) and therefore lie in the plane of the paper. The first is perpendicular to $\boldsymbol{\rho}$ and the second to $\dot{\boldsymbol{\rho}}_{\mathrm{r}}$. The velocity and acceleration components are shown in Fig. 5.19. Of the four sketches (a) shows the components of velocity and (b) the two components of acceleration of a point on the track. (c) shows the acceleration of P relative to the track; this has two components, one centripetal (a v^2/R term) arising from the curvature of the track and one tangential having magnitude $\frac{d}{dt}(\dot{\rho}_{\mathrm{r}})$. (d) shows the Coriolis component, which the law of

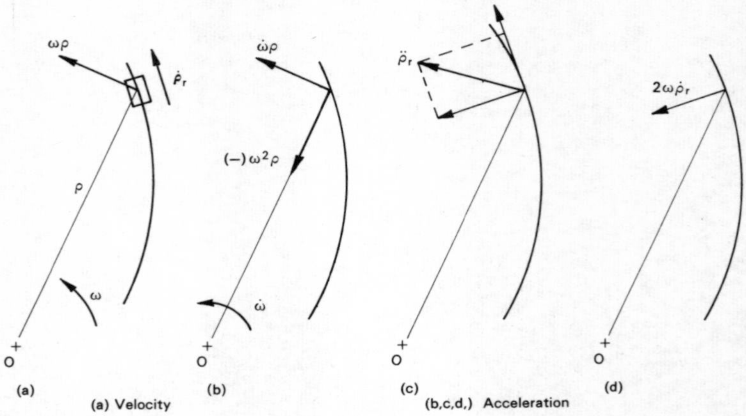

FIG. 5. 19

vector multiplication tells us is perpendicular to both $\boldsymbol{\omega}$ and $\dot{\boldsymbol{\rho}}_r$, its direction can most reliably be found from the right-hand screw rule.

As a complete example of the two-dimensional case consider the Geneva mechanism in Fig. 5. 20 of the type used for imparting interrupted motion in such devices as cine projectors and wrapping and stamping machines. The projecting pin B is fixed to a wheel rotating clockwise with a constant angular velocit about the fixed axis A. The pin engages the slotted wheel which rotates about a parallel fixed axis C. When the pin is so engaged the slotted wheel rotates; when the pin is not engaged the slotted wheel is held stationary by the engagement of the rim R in the curves between the slots. To design the various bits and pieces in the machine, the designer would need to know the accelerations of the rotating parts throughout the cycle of operations so that he could calculate the forces. Whether the values are to be found analytically or graphically, it is helpful to draw velocity and acceleration diagrams. Consider a case in which the dimensions are AB = 50 mm and AC = 100 mm, the wheel on axis A is running at constant angular velocity 10 rad/s. We will draw the diagrams for the position when angle CAB is 30°.

To find the velocity and acceleration of the slotted wheel the pin is considered from two points of view. One is simply to regard B as a point on whee A which is moving with constant angular velocity; it is therefore easy to calculate the velocity and acceleration of B. The other is to treat B as a point which is mov ing along a track on wheel C which is itself rotating. Figure 5. 21 shows the position of the mechanism (a), the velocity diagram (b) and the acceleration diagram (c In the position diagram it is useful to describe the point B in two ways: as B on AI and as D on CD; these are the P and Q of our general analysis, coincident at the instant considered, but moving one past the other. The velocity diagram is easy tc draw; the magnitude of the velocity of B can be calculated and the direction is

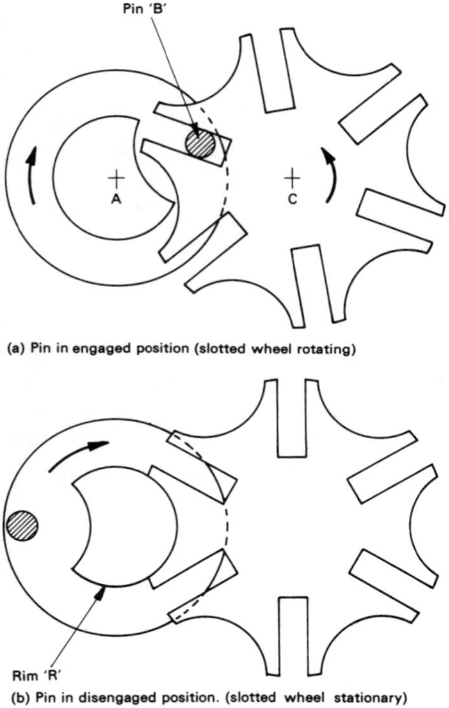

FIG. 5. 20 The Geneva mechanism

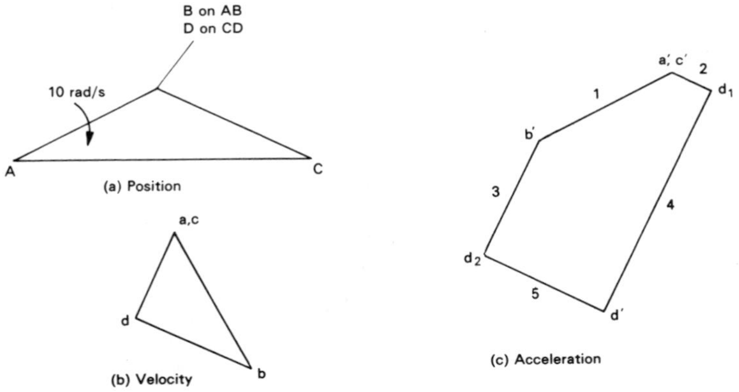

FIG. 5. 21 Velocity and acceleration diagrams for the Geneva mechanism

known, so ab can be drawn first; the direction of the velocity of D(ad) must be perpendicular to CD so this *direction* can be drawn next; finally the velocity of B relative to D(db) must be directed along CD, so this *direction* can be drawn, and the intersection of the two directions gives the point d.

The acceleration diagram is not quite so straightforward. Figure 5.21 (c) shows what we want to end up with. a′b′ can be put in straight away; that is the acceleration of B calculated by regarding it as a point on wheel A. It is centripetal only and is simply $r\omega^2$, in the direction BA. The remainder of the diagram is made up of the various components of the acceleration of B regarded as a point moving along the track DC on wheel C. $c'd_1$ and d_1d' are the centripetal and transverse components of the acceleration of the point D fixed to wheel C; the centripetal component can be calculated from the information on the velocity diagram and equals $\frac{(cd)^2}{CD}$, the transverse component cannot be calculated from the velocities but its direction is known to be perpendicular to CD. $d'd_2$ is the (sliding) acceleration of B relative to D; its magnitude is not yet known but its direction is plainly along CD (one way or the other). Finally d_2b' is the Coriolis component $2\dot{\theta}u$, where $\dot{\theta}$ is the angular velocity of wheel C (equal to $\frac{cd}{CD}$) and u is the sliding velocity of B along DC (equal to db). Summarising the accelerations:

(1) a′b′ - the absolute acceleration of B
(2) $c'd_1$ - centripetal } components of acceleration of point D on
(4) d_1d' - transverse } moving frame
(5) $d'd_2$ - acceleration of B relative to D, sliding only in this case, because the track is straight
(3) d_2b' - the Coriolis component

The numbers in front of the terms show the order in which the lines could be accurately drawn in a graphical solution: (2)-(5) must of course add together vectorially to equal (1). (1), (2) and (3) can be calculated from velocities, either given or found from the velocity diagram, (4) and (5) must be found from the intersection of the known directions.

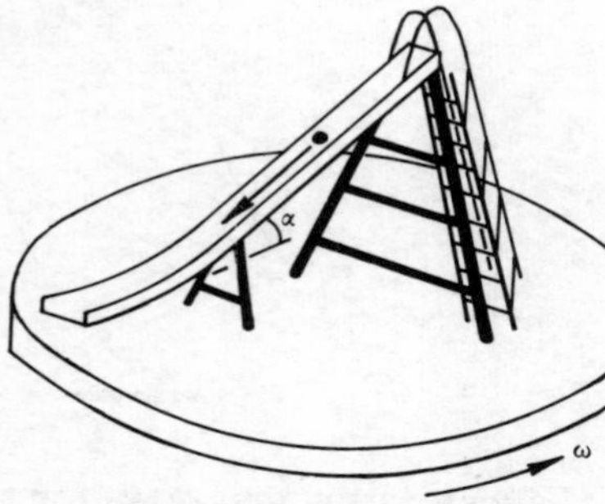

FIG. 5.22

With the acceleration diagram complete, the angular acceleration of the wheel C, $\ddot{\theta}$, can now be calculated. It is the transverse component of acceleration of D (an $\boldsymbol{r}\ddot{\boldsymbol{\theta}}$ term) divided by the length CD, i.e. d_1d'/CD.

It is now time to look at an example in three dimensions. Impressed by your earlier success on the roundabout, and prevailing still further on the fairground proprietor's tolerance, suppose you now set up a simple playground slide on the roundabout, as in Fig. 5. 22. Write down what you think will be the components of the acceleration of a child sliding down the slide, and then see if you agree with the values shown on Fig. 5. 23. Take ω = constant and regard the child as a particle.

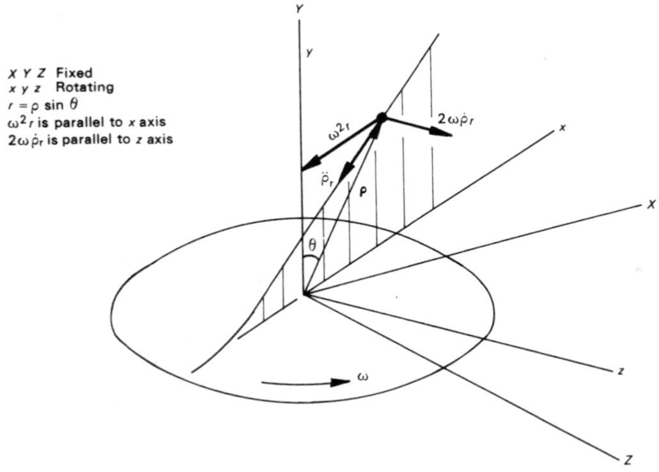

FIG. 5. 23

This chapter has been concerned with motion (velocity and acceleration), not forces. We have found out how to calculate (in principle), or estimate graphically, the details of the motion of simple and some not-so-simple mechanisms.

Bibliography

Morrison, J. L. M., and Crossland, B., *An Introduction to the Mechanics of Machines*, Longmans, 1964

Examples

In some of the examples below, you may prefer a graphical to an analytical method.

Ex. 5.1 The diagram shows a sketch of a mechanism in which the link BC, pivoted at C, oscillates about the vertical position. The end B is free to turn about a pivot on the slider, which can move up and down the link AB.

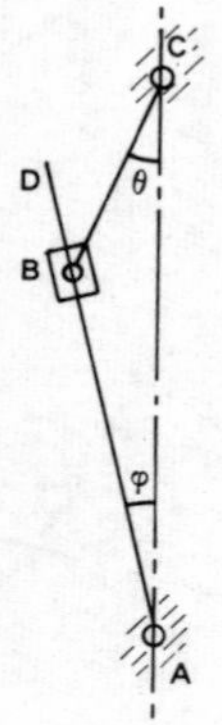

Ex. 5.1

If ω is the angular velocity of link BC, and Ω the angular velocity of link AB, sketch a velocity diagram and hence show that

$$\frac{\Omega}{\omega} = \frac{\sin \phi}{\sin \theta} \cos (\theta + \phi)$$

Ex. 5.2 A crank OP and connecting rod PQ are arranged so that the piston slides on an axis passing through O. PQ is 1 m, OP is 0.25 m. Find the velocity and acceleration of the piston when the angle POQ is 60°, if the crank is rotating at a uniform angular velocity of 10 rad/s. [2.4 m/s, 9.4 m/s²]

Ex. 5.3 Describe and justify a simple approximate method for calculating the maximum piston velocity in a reciprocating engine running at constant crank speed.

A petrol engine with a crank radius of 4 cm and a connecting rod length of 14 cm is running at 2000 rev/min. What will be the greatest piston velocity and at what point in the stroke will it occur? [87 m/s at 3.4 cm from t.d.c.]

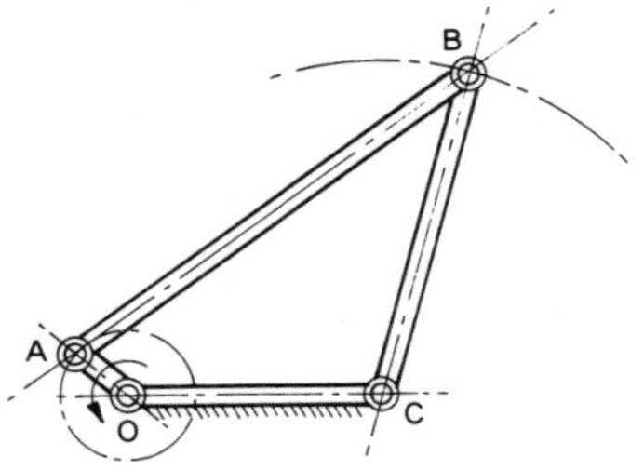

Ex. 5. 4

Ex. 5. 4 The mechanism shown consists of four rigid links, pin-jointed: OA 3cm, AB 23 cm, BC 16 cm, OC 12 cm. Link OC is fixed and OA rotates with a constant angular velocity of 20 rad/s.

Estimate the angle between the extreme positions of the travel of the oscillating link BC, and its angular acceleration at both extremes.

[46°, 110 and 267 rad/s^2]

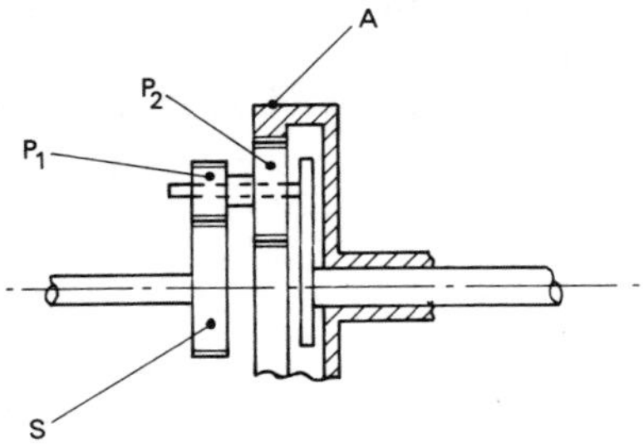

Ex. 5. 5

Ex. 55 In the epicyclic gear shown the numbers of teeth on the wheels are: sun 30, annulus 80, compound planet P_1 20 and P_2 40. What will be the gear ratio with the annulus fixed? [7 : 3]

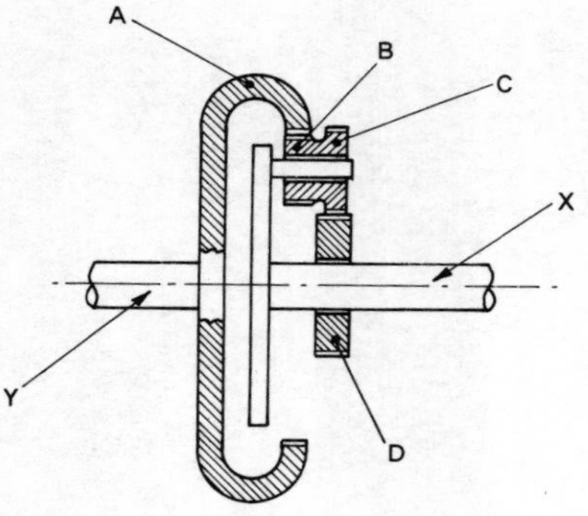

Ex. 5.6

Ex. 5.6 In the epicyclic gear shown the numbers of teeth on the wheels are: annulus A = 60, sun D = 25, compound planet B = 15, C = 20. The shaft X is free to rotate relative to the sun wheel D, the annulus is keyed to the shaft Y and the spider carrying the planets is keyed to the shaft X.

(a) If the input to shaft X is 50 kW at 250 rad/s and D is held stationary, find the output speed and torque from shaft Y.

(b) If with the same input to X, D now rotates in the same direction as X at 50 rad/s, find the output speed and power from shaft Y and the necessary torque on D. Neglect friction throughout.

Notes: (a) Power = rate of work measured in N m/s, i.e. watts.

(b) Because there are no accelerations, the *torques* will be the same in part (b) as in part (a); they will satisfy equilibrium.

[328 rad/s, 152 N m; 312.5 rad/s, 47.6 kW, 48 N m braking]

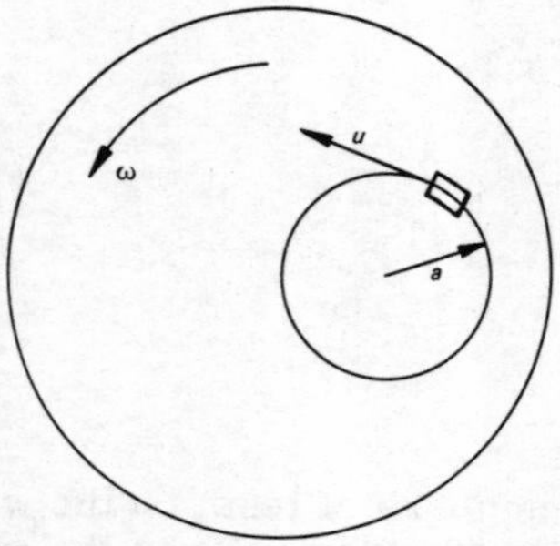

Ex. 5.7

Ex. 5. 7 A child sets up his model train on a flat level roundabout, with a point on the circular track coincident with the centre of rotation of the roundabout.

Show on a sketch the components of acceleration of a clockwork engine moving with constant speed u anticlockwise relative the track, when the roundabout is rotating at constant angular velocity ω anticlockwise.

If the mass of the engine is 1 kg and the radius of the track is 0. 4 m, find the maximum horizontal force between the engine and the rails when u is 1 m/s and ω is 2 rad/s. [9. 7 N]

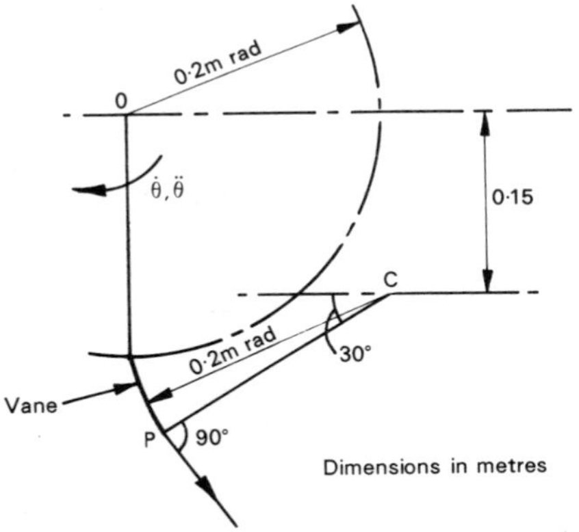

Ex. 5. 8

Ex. 5. 8 The figure shows one vane of an impellor wheel, the centre of curvature of the circular vane being at that instant at C. A fluid element P at the tip of the vane has velocity and acceleration, relative to the vane, 0. 5 m/s and 1 m/s^2 in the direction shown.

Find the velocity and acceleration of P relative to O when the impellor wheel has $\dot{\theta} = 2$ rad/s and $\ddot{\theta} = 2$ rad/s^2 in the sense shown. [0. 54 m/s, 0. 78 m/s^2]

Ex. 5. 9 At the equator a projectile is fired vertically upwards. It reaches a height h and falls down again. Neglecting air resistance, show that the distance between the firing point and the point where the projectile falls is

$$\frac{8\Omega}{3}\sqrt{\frac{2h^3}{g}}$$

where Ω is the earth's angular velocity.

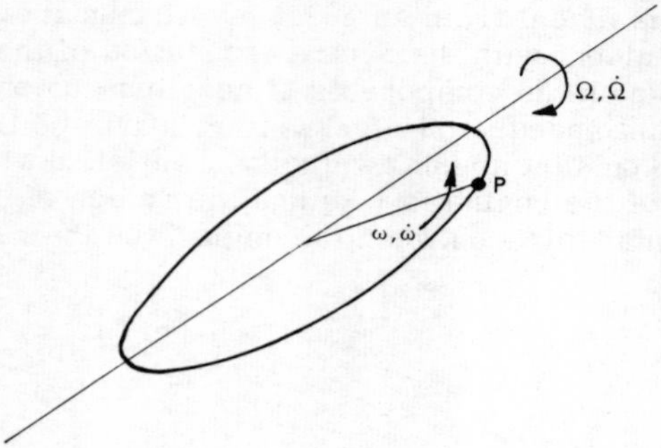

Ex. 5. 10

Ex. 5. 10 A point P moves around a circular track at ω, $\dot{\omega}$. The track itself rotates about a diameter at Ω, $\dot{\Omega}$.

On a sketch show all the components of acceleration of P. (Remember: acceleration of Q, a point on the track instantaneously coincident with P + acceleration of P relative to Q + Coriolis)

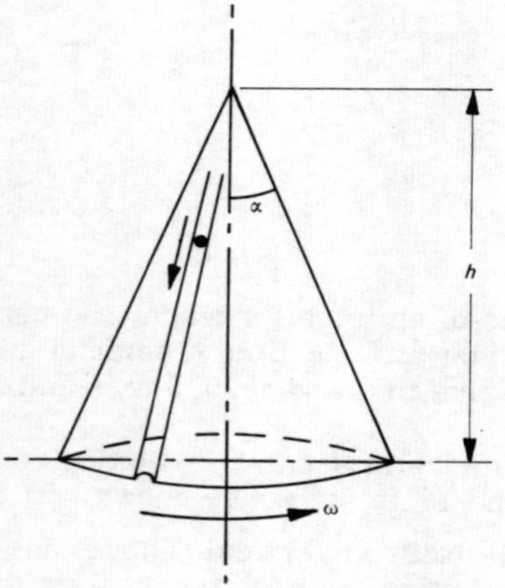

Ex. 5. 11

Ex. 5. 11 A modified roulette wheel could be made on which the usual rotating wheel with numbered slots remained stationary, and the small ball was distributed at random by rolling it down a groove cut in the side of a rotating cone.

Treating the ball as a particle, write down the equations of motion of the ball when it is rolling down the groove, and the cone (of height h and semi-angle α) is rotating at constant angular velocity ω. Restrict your attention to the time before the ball 'flies off' the cone, but write down an expression for the condition that the ball will stay on the cone until it reaches the bottom.

[$h\omega^2 \leqslant g$]

6  *Particle Dynamics*

The applications of particle dynamics are very limited in engineering, because most common engineering devices involve the rotation of solid bodies. Perhaps the most numerous are in situations where large solid bodies are constrained to move in translation only, with no rotation. We shall see in the next chapter that these can in some respects be treated as particles. For some simple considerations, road vehicles and railway trucks and carriages can be so treated. There are certainly enough applications to make the study worthwhile in itself, but another good reason for studying particle dynamics is that it brings home many important ideas.

We shall first apply Newton's second law in the straightforward way, but we shall see that it becomes a much more powerful tool when it is integrated. Integrated with respect to time it leads to impulse and momentum; integrated with respect to displacement it leads to work and energy. The motion of individual particles under applied forces, and the interactions between particles, will be examined in detail.

We have written Newton's second law

$$\mathbf{F} = \frac{\mathrm{d}}{\mathrm{d}t}(m\mathbf{v}) \text{ and } \mathbf{F} = m\mathbf{a}$$

The second of these, in the notation of Chapter 5, is

$$\mathbf{F} = m\ddot{\mathbf{r}}$$

which can in turn be written in cartesian components

$$F_x = m\ddot{x}$$
$$F_y = m\ddot{y}$$
$$F_z = m\ddot{z}$$

or of course in any other coordinate system.

Integration of these in a particular context will yield, in principle, velocity and displacement under any set of applied forces.

For a start, consider a very simple one-dimensional example. Suppose you are a member of a well-established gliding club which launches its powerless crafts by giving them a good push off the side of a cliff, where they

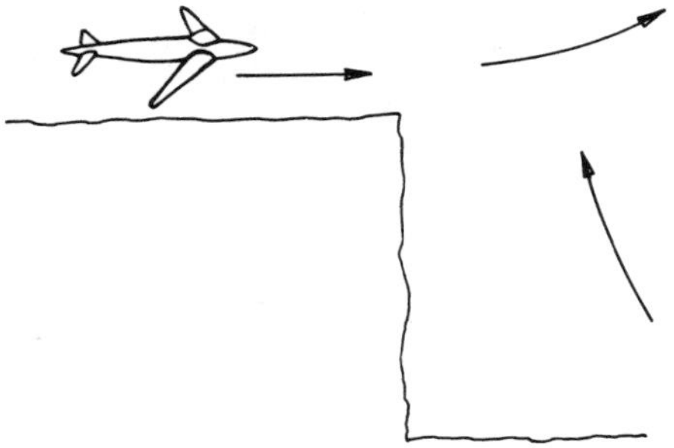

FIG. 6. 1

can take advantage of the up-currents of air, Fig. 6. 1. The old winch and tow-rope used to give the push are due for replacement, and you are given the job. Being an engineer, you consider other ways of doing the business. Two of the possibilities you examine are:

1. a simple and light, slowly-burning rocket motor which would permanently be fixed to each glider and would give a constant thrust P for a time T; and
2. a catapult trolley on which each glider would be placed for the launch, to be driven by a linear electric motor giving a constant thrust P over a length of track, s.

After deciding on a safe value of P for the gliders, you will want to know the necessary burning time of the rocket, and the minimum length of linear motor track, to give the desired launching velocity V.

For this one-dimensional case we can write

$$P = m\,\frac{\mathrm{d}^2x}{\mathrm{d}t^2} = m\,\frac{\mathrm{d}v}{\mathrm{d}t}$$

$$\int_0^T P\mathrm{d}t = \int_0^T m\,\frac{\mathrm{d}v}{\mathrm{d}t}\,\mathrm{d}t$$

$$\Big[Pt\Big]_0^T = \Big[mv\Big]_0^V$$

which gives the velocity V after time T under the constant force P directly:

$$PT = mV$$

which is the relation you want for the rocket. (Note that we have assumed that the mass of propellant ejected by the rocket is negligible compared with the mass of the glider, so that m can be regarded as constant. Later in this chapter we shall examine examples of rocketry where this is not permissible.)

For the linear motor we could integrate again with respect to time and solve for the track length in terms of the launching velocity. But as you well know we can proceed directly to the answer you want by the old trick of disposing of t as the variable in the expression for acceleration.

$$\frac{d^2x}{dt^2} = \frac{dv}{dt} = \frac{dv}{dx}\frac{dx}{dt} = v\frac{dv}{dx}$$

then $P = mv\dfrac{dv}{dx}$

$$\int_0^s P\,dx = m\int_0^s v\frac{dv}{dx}\,dx$$

which gives

$$Ps = \frac{mV^2}{2}$$

The two expressions in P and V would enable you to take the next step in the design.

This simple example points the way to two important integrated forms of Newton's second law, one with repsect to time, the other with respect to displacement.

The Time Integral. Impulse and Momentum

$$\mathbf{F} = m\ddot{\mathbf{r}} = m\frac{d\mathbf{v}}{dt}$$

Integrating both sides with respect to t, for m = constant,

$$\int_{t_1}^{t_2} \mathbf{F}\,dt = m\mathbf{v}_2 - m\mathbf{v}_1 \tag{6.1}$$

The vector $m\mathbf{v}$ is called the linear momentum or simply the *momentum* of the particle m and the integral term on the left is called the *impulse* of the force $\mathbf{F}$. The result is therefore that the time integral of the force, or impulse, is equal to the change of momentum of the particle. Equation (6.1) is a vector equation and can be written in components in any appropriate coordinate system. For example

$$\int_{t_1}^{t_2} F_x\,dt = m\dot{x}_2 - m\dot{x}_1, \text{ etc.} \tag{6.2}$$

It is plain that if no force is applied to a particle, its momentum will not change, but of course that is exactly what the second law says. Equation (6.2) goes further, saying that if the applied forces have a zero component in any direction there will be no momentum change in that direction, but that also is quite clear in the original statement of the second law.

A much more important consequence of this is that if a *system of particles* is subjected to no forces *external to the system*, the total momentum of the system will be constant. Whatever forces there may be between the particles in the system, whether they be due to collisions or gravitation or pieces of string, they will all act in equal and opposite pairs by Newton's third law, and so for every $\int \mathbf{F}\,dt$ there will be an $\int -\mathbf{F}\,dt$; more of this later.

It should be noted that some writers use the term 'impulse' to apply only to the integral of a very large force acting for a very short time; viz. impact conditions. In such conditions it may be very difficult to measure the force and its variation with time during the short interval $t_2 - t_1$, but relatively easy to measure the impulse by measuring directly the change in $m\mathbf{v}$. For example it would be quite easy with instruments now available to measure the velocity of a golf ball as it leaves the tee, but not nearly so easy to measure the force between the driver and the ball. In calculations to do with this sort of thing it is usually in order to neglect non-impact forces, such as the weight of the golf ball, compared with the impact forces *during the period of the impact*.

In this book the term impulse is used simply as the time integral of the force; the short-term forces of impact conditions will be called impact forces.

The Displacement Integral. Work and Energy

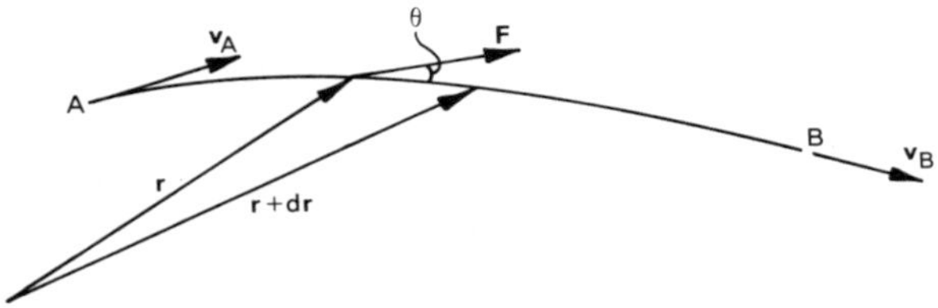

FIG. 6. 2

Consider the motion in Fig. 6.2 of a particle along an arbitrary path between points A and B. Suppose its velocity at those two points is $\mathbf{v}_A$ and $\mathbf{v}_B$, and it is acted on by varying forces whose resultant at any time is $\mathbf{F}$:

$$\mathbf{F} = m\,\frac{d\mathbf{v}}{dt}$$

Knowing from earlier considerations that work and energy are scalar quantities, we start the integration by taking the scalar product of both sides with an increment of displacement dr

$$\mathbf{F} \cdot \mathrm{d}\mathbf{r} = m \frac{\mathrm{d}\mathbf{v}}{\mathrm{d}t} \cdot \mathrm{d}\mathbf{r}$$

The left-hand side is the work done by the force **F** as its point of application moves through the displacement **dr**. Its magnitude is

$$\mathbf{F} \cdot \mathrm{d}\mathbf{r} = F \cos\theta \, \mathrm{d}s$$

where θ is the angle between **F** and the path, and ds is the incremental length of path between **r** and **r** + **dr**. The work done on the particle between A and B is therefore

$$W = \int_A^B \mathbf{F} \cdot \mathrm{d}\mathbf{r} = \int_A^B F \cos\theta \, \mathrm{d}s \qquad (6.3)$$

This integral is called a line integral, and its value may or may not depend upon the path followed from A to B, according to the nature of **F**. We shall have another look at that in a moment, but first we must integrate the right-hand side of our equation. $\mathrm{d}\mathbf{r} = \mathbf{v}\mathrm{d}t$, so the right-hand side becomes

$$m \frac{\mathrm{d}\mathbf{v}}{\mathrm{d}t} \cdot \mathrm{d}\mathbf{r} = m \frac{\mathrm{d}\mathbf{v}}{\mathrm{d}t} \cdot \mathbf{v}\mathrm{d}t = m\mathbf{v} \cdot \mathrm{d}\mathbf{v}$$

$$= \tfrac{1}{2} m\mathrm{d}(\mathbf{v} \cdot \mathbf{v}) = \tfrac{1}{2} m\mathrm{d}(v^2)$$

Integrating this and equating the two sides

$$\int_A^B \mathbf{F} \cdot \mathrm{d}\mathbf{r} = \tfrac{1}{2} mv_B^2 - \tfrac{1}{2} mv_A^2 \qquad (6.4)$$

The quantity $\tfrac{1}{2} mv^2$ is called the kinetic energy T of the mass m. So we have the general result that the work done by the resultant **F** of all the forces acting on the particle m is equal to the increase of kinetic energy of the particle.

Clearly the term **F** . dr can be positive or negative, according to whether the force component and displacement are in the same or opposite direction.

Returning now to the value of $\int_A^B \mathbf{F} \cdot \mathrm{d}\mathbf{r}$, let us look first at some examples that we already know about. In the glider launched by a catapult the value of the integral was simply Fs, where F was a constant force acting in the direction of motion and throughout the distance s. Another situation in which the force would be constant and everywhere in the same line of action as the motion, but *opposite* in sense, is that of a particle sliding over a surface with a constant coefficient of friction. If the surface is horizontal and there are no applied forces, friction will be the only work-doing force. Whatever the length or shape of the path the work done by the friction force on the particle, i.e. the $\int \mathbf{F} \cdot \mathrm{d}\mathbf{r}$ will be simply the constant force F times the length of the path s. It will of course be negative and therefore the kinetic energy of the particle will be reduced.

In both of these examples, the lines of action of the force and the displacement were always coincident; cos θ in (6.3) equal to $+1$ or -1. The other extreme is when the force and displacement are perpendicular, $\cos\theta = 0$ and the work done by the force is zero.

If we turn our thoughts back to Chapter 4 for a moment, we recall writing equations for static situations which said in effect 'the work done by the external forces equals the strain energy stored by the elastic structure'. How do we now reconcile those statements with the work–kinetic energy equation (6.4)? Thinking of any part of the structure as a particle, Chapter 4 was concerned only with static, i.e. equilibrium, situations, where all the forces on any particle had a resultant of zero, so equation (6.4) would simply be zero on both sides. We must clearly bring all these ideas together.

This present chapter is about single particles, which structures certainly are not, but to get our ideas straight we can *conceptually* separate kinetic energy and strain energy completely by thinking for the present in terms of particles with mass, and elastic springs *without* mass. Consider the arrangement in Fig. 6.3 where a particle m with velocity v_A approaches an elastic spring, Fig. 6.3 (a). When the particle meets the spring, the spring will start to be compressed and

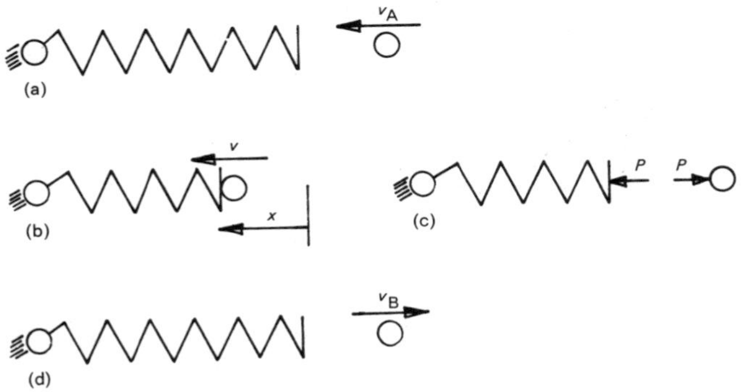

FIG. 6.3

the particle will start to be retarded, Fig. 6.3 (b). At any stage during the compression, we can conceptually isolate the particle from the spring provided we imagine the internal force P between the two as opposite external forces, Fig. 6.3 (c). The work done on the spring is $\int P\mathrm{d}x$ and the work done on the particle is $\int -P\,\mathrm{d}x$; in other words the work done *by* the particle on the spring is $\int +P\mathrm{d}x$. There is an exchange between kinetic energy in the particle and strain energy in the spring—the spring has no kinetic energy at any stage because it has no mass, remember.

When the particle comes to rest the whole of its kinetic energy has been transferred to the spring, and the force P has reached its maximum value. The particle starts then to move from left to right and will eventually leave the spring with velocity v_B, as in Fig. 6. 3 (d).

The difference between this example and the friction one above is that the friction force always opposed the motion of the mass, whereas the spring force in Fig. 6. 3 always acts from left to right on the particle, whether it is travelling leftwards or rightwards. So the spring opposes the motion of approach, but promotes the motion of departure; in other words, kinetic energy is transferred to the spring as strain energy until the velocity of the mass is zero, and then *returned* to the mass as kinetic energy as it moves off to the right. If the spring is

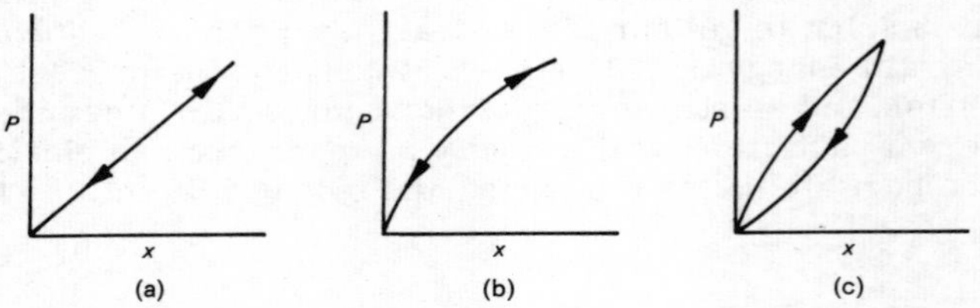

FIG. 6. 4

a perfectly elastic spring as in Fig. 6. 4 (a) or (b) (it doesn't have to be linear), the whole of the original kinetic energy of the mass will be restored, and $v_B = v_A$. In a real spring the recovery will not be quite complete. The 'loss' of energy is shown by the 'elastic hysteresis' loop of Fig. 6. 4 (c); for metals of the sort used for springs this loop is extremely thin, so the hysteresis loss is tiny.

What happens to the 'loss' of energy, and for that matter to the whole of the kinetic energy which is lost in the friction example above? It is transferred to the material of the spring or the friction surface and raises the temperature of the material. But unlike the elastic strain energy it is not recoverable, except by the use of an elaborate 'heat engine', and then only in part (see [1] and the deadly work of the 'other' second law).

How would the events just described be modified if the spring were made of plasticine? If you are not sure, try it. Having done the experiment what have you concluded about the stress/strain curve for plasticine?

Potential energy

In these simple examples we have seen that work done by external forces can change the kinetic energy of a particle or the strain energy of a spring, and that these last two can be reversibly interchanged in the special case of a perfectly elastic spring. A perfectly elastic spring force is an example of a so-called *conservative force*, and the strain energy stored by the spring is an example of *potential energy*. Friction force on the other hand is an example of a *dissipative force*.

The Principle of Conservation of Mechanical Energy says that for a particle acted upon only by conservative forces the sum of the potential energy and kinetic energy is constant. This is getting dangerously close to 'energy is conserved in systems where energy is conserved'. Nevertheless, if there is not a significant dissipative element present, the conservation principle is very useful, either in the form just stated or in the more general work—energy form:

The work done on a particle by the externally applied forces equals the change in the sum of the potential and kinetic energies. (6.5)

(on the understanding that potential energy includes elastic strain energy).

The constancy of (kinetic energy + potential energy) in the absence of externally applied forces, i.e. the conservation of mechanical energy, can only be an *idealisation* of real engineering systems. Even in the best-kept mechanical systems there are always unwanted 'applied forces' in the form of friction of various sorts, and they cause losses of mechanical energy all over the place. The engineer does all he can to minimise the losses, partly for economic reasons and partly to avoid the whole thing going up in smoke—remember that the 'losses' are simply temperature-raising energy transfers.

The other important conservative force in engineering mechanics is that due to gravity. It is a characteristic of, indeed the definition of, a conservative force that the work done by or against it depends only on the initial and final positions of its point of application, and not upon the path taken between those positions.* In Fig. 6. 5 consider the work done in moving a particle m from $\mathbf{r}_1$ to

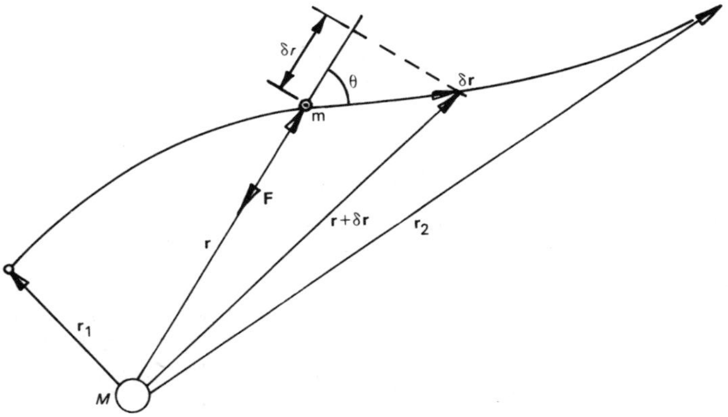

FIG. 6. 5

* It can be shown that any central force field whose magnitude is a function only of radius is conservative.

$\mathbf{r}_2$ in the gravitational field of M. From (3.2) the gravitational force on m is

$$\mathbf{F} = \frac{GMm}{r^2}(-\hat{\mathbf{r}})$$

where $\hat{\mathbf{r}}$ is the radial unit vector. **F** is shown in the $-\hat{\mathbf{r}}$ direction in the diagram. The work done by **F** over an increment of path $\mathbf{dr}$ is

$$dW = \mathbf{F}\,.\,\mathbf{dr} = -|\mathbf{F}||\mathbf{dr}|\cos\theta$$

where, as before, $|\mathbf{F}|$ signifies the magnitude of **F**. In the diagram it can be seen that

$$|\mathbf{dr}|\cos\theta = dr$$

so

$$dW = -F dr = -\frac{GMm}{r^2}dr$$

This is a scalar expression and the only variable is r; the limits of integration are therefore r_1 and r_2. The value of the integral does not depend on the route taken from $\mathbf{r}_1$ to $\mathbf{r}_2$, nor indeed for this radial force field upon the directions of $\mathbf{r}_1$ and $\mathbf{r}_2$ but only on their scalar values r_1 and r_2. Integrating,

$$\int_{r_1}^{r_2} dW = -GMm\int_{r_1}^{r_2}\frac{dr}{r^2} = GMm\left(\frac{1}{r_2} - \frac{1}{r_1}\right)$$

If $r_2 > r_1$ then the work done by **F** is negative, because the force and displacement are in opposite senses. The work which must be done *on m* to increase its distance from M is *defined* as the increase in *potential energy V* of m.

Thus $$V_2 - V_1 = GMm\left(\frac{1}{r_1} - \frac{1}{r_2}\right) \qquad (6.6)$$

We should note here that the term 'potential energy' is commonly used to describe gravitational potential energy, while the potential energy stored in an elastic spring or structure is usually called its 'strain energy'.

As with elastic strain energy, the work which must be done to change the potential energy of a particle could be done by or against an external force, or be transferred to or from the kinetic energy of the particle, or by a combination of both. The general statement (6.5) describes the work and energy transfers, with the term potential energy embracing gravitational potential energy and elastic strain energy.

It is customary to take a datum value of r_1 and set $V_1 = 0$ there; then $V = V_2$ is the value of the potential energy at $r = r_2$. Except in earth-bound calculations, where the changes in r are negligible compared with the radius of the earth (a special case that we shall examine in a moment), the datum is taken at infinity; i.e. $V_1 = 0$ at $r_1 = \infty$ and

$$V = -\frac{GMm}{r} \tag{6.7}$$

which is everywhere negative, but gets less (by getting more netagive) as r gets smaller. There is no physical significance in (6.7) being negative at all r. If the world's airlines decided to use the top of Everest as the datum for altitude, all aircraft for some of their flight-time, and many for all of it, would be flying at negative altitude; all pilots would know that when they were flying at positive altitude there was no danger of flying into a mountain.

The expression (6.7) can be applied to the earth, for values of r greater than the earth's radius, if we are satisfied to represent the earth as a sphere with radial symmetry and mass M. The form (6.7) must be used if we are concerned with changes in r which are significant compared with r itself (space-flight); but for situations where the changes in r are negligible compared with r, the expression can be simplified.

$$V_2 - V_1 = GMm\left(\frac{1}{r_1} - \frac{1}{r_2}\right)$$

If $r_2 = r_1 + \Delta$, where $\Delta << r_1$

$$V_2 - V_1 = GMm\left(\frac{1}{r_1} - \frac{1}{r_1 + \Delta}\right)$$

$$\frac{1}{r_1 + \Delta} = \frac{1}{r_1\left(1 + \dfrac{\Delta}{r_1}\right)} \approx \frac{1}{r_1}\left(1 - \frac{\Delta}{r_1}\right)$$

So $$\frac{1}{r_1} - \frac{1}{r_1 + \Delta} = \frac{1}{r_1} - \frac{1}{r_1}\left(1 - \frac{\Delta}{r_1}\right) = \frac{\Delta}{r_1^2}$$

and $$V_2 - V_1 = GMm\frac{\Delta}{r_1^2}$$

Now from (3.3) we know that the local value of g, the acceleration due to gravity, is

$$g = \frac{GM}{r^2}$$

Therefore

$$V_2 - V_1 = mg\Delta \tag{6.8}$$

where g has the value appropriate to r_1. So for small changes of r, i.e. changes of 'height' Δ, the change in potential energy is simply $mg\Delta$, in words the weight times the change of height. The datum is taken at any convenient level to suit the particular situation; its actual value is irrelevant because it is only *changes* of potential energy that matter. Rember that $mg\Delta$ can only be used when $\Delta << r$, where r is the distance from the centre of mass of the central body, e.g. the earth.

Summary

We have come a long way since we pushed the glider off the cliff, so it would be useful to summarise the results for particles so far.

The time integral of Newton's Second Law gives the relation between impulse and momentum change. In the absence of external forces there is no change of momentum—particularly useful in calculations involving velocities and *times*.

The displacement integral of Newton's Second Law gives the work—energy relation 'The work done on a particle by the externally applied forces equals the change in the sum of the potential and kinetic energies'—particularly useful in calculations involving velocities and *displacements*.

Gravitational and perfectly elastic forces are conservative. Friction is dissipative. Gravitational potential energy changes depend only on changes of distance from the attracting mass; for small changes, the potential energy change is $mg\Delta$, usually written as mgh.

These principles will be illustrated in a number of examples below, mainly in a study of a simple collision and in an elementary examination of satellite orbits. But at this point it is worth drawing attention to an important distinction between momentum and energy considerations in mechanical systems, of particular importance when we consider not a single particle but a number of particles or real bodies.

When energy conservation is applied to mechanical systems, it can never be better than in idealisation. There is much humming and hawing about 'perfect elastic solids', 'absence of dissipative forces' and the like, because when applied to *mechanical* energy, the conservation principle is being abused by the omission of non-mechanical terms, mainly thermodynamic.

The application of momentum conservation on the other hand, or the impulse—momentum relation, is fundamentally sound. To apply it, a collision for example need not be between two 'perfectly elastic spheres'; momentum is conserved in a collision between a loaf of bread and a jar of jam (even if the jar breaks), but of course kinetic energy is not.

Interactions between particles

Recalling the integrations of Newton's Second Law for straight-line motion:

Time	*Displacement*
$F = m\dfrac{dv}{dt}$	$F = mv\dfrac{dv}{dx}$
$\int_{t_1}^{t_2} F\,dt = mv_2 - mv_1$	$\int_{x_1}^{x_2} F\,dx = \dfrac{mv_2^2}{2} - \dfrac{mv_1^2}{2}$

If two particles sustain the mutual forces of an interaction, the forces will be equal and opposite, whatever their nature, by Newton's Third Law, at all times during the interaction. If there are no additional external forces, $\int F\,dt$ will be the same for both except for sign; the $\int F\,dt$ of one particle and the $\int -F\,dt$ of the other will be exactly equal and opposite. The momentum surrendered by one particle will be gained by the other.

This conservation of momentum may be accompanied by an energy loss or an energy *gain*, from an internal supply of energy; for example due to the burning of propellant in a gun. Sir Charles Inglis is worth quoting on this [2]: 'Take for example, a projectile fired from a gun. If questioned, most students will glibly and correctly answer that the forward momentum given to the projectile is equal to the backward momentum given to the gun; but, if pressed to explain why the kinetic energies are different, the clarity of their replies often leaves much to be desired. But, if the distinction between the time effect of a force and its space effect is appreciated, the explanation is quite obvious. The force on the projectile and the force on the gun are equal and opposite, and they act for the same time. The time effects of the forces on the projectile and the gun are equal and opposite, and the momentum given to them must also be equal and opposite. The forces, however, do not act through the same distance; the space integral for the force on the projectile is many times greater than the space integral of the equal force acting on the gun, consequently the kinetic energy given to the projectile greatly exceeds that given to the gun, which is a merciful safeguard for those who have to handle these weapons of destruction'.

The same point is made in symbols in Fig. 6.6.

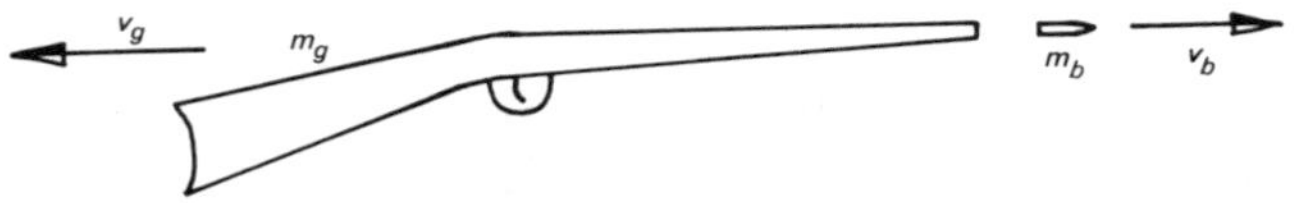

FIG. 6.6

$$\text{Momentum:} \qquad m_g v_g = m_b v_b$$

$$\text{Kinetic energy of gun} = \tfrac{1}{2}\, m_g v_g^2$$

$$\text{Kinetic energy of bullet} = \tfrac{1}{2}\, m_b v_b^2$$

$$\frac{\text{kinetic energy of gun}}{\text{kinetic energy of bullet}} = \frac{\tfrac{1}{2}\, \cancel{m_g v_g} \times v_g}{\tfrac{1}{2}\, \cancel{m_b v_b} \times v_b} = \frac{m_b}{m_g},$$

a ratio which is very small indeed. Of course this simple sum ignores the propellant gases.

An old-fashioned method of measuring the velocity of the bullet employs both principles, momentum and energy. It is called the ballistic pendulum and is shown in Fig. 6. 7, which explains itself. The bullet is aimed at the centre

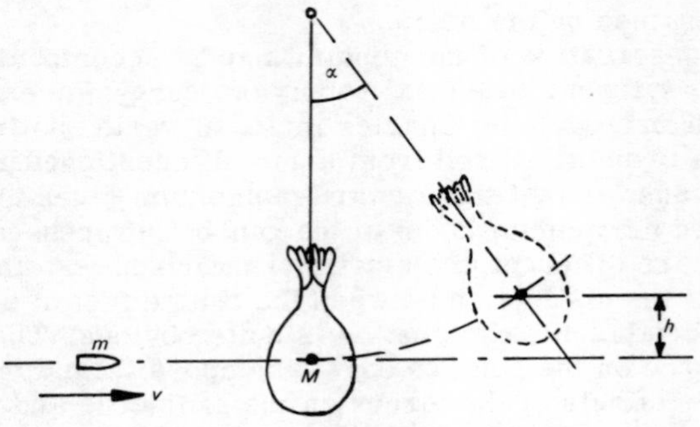

FIG. 6. 7 The ballistic pendulum

of mass of a sand-bag which hangs freely on a string. The bag swings away from the vertical, the angle α is the measurement, and the rise of the centre of mass, h, can be calculated. The conditions are impact conditions in that the bag will hardly have moved when the bullet becomes embedded. The velocity of the bullet and bag can therefore be calculated on the basis of an instantaneous momentum transfer; $mv \rightarrow (M + m)V$, which in the absence of external forces (in the direction of v) gives

$$mv = (M + m)V \text{ ... momentum}$$

V can be calculated from h using the work—energy relation. There is no work done by the external force (the supporting force at the pivot holding the string), so the change in the sum of the potential and kinetic energies is zero,

$$(m + M)\frac{V^2}{2} = (m + M)gh$$

so

$$V^2 = 2gh \text{ ... energy}$$

and we already have from momentum

$$v = \left(\frac{M + m}{m}\right)V = \frac{M + m}{m}\sqrt{2gh}$$

v can therefore be calculated, albeit approximately.

You might have been tempted to take the short cut of writing $\frac{1}{2}mv^2 = (M + m)gh$, applying energy conservation to the overall situation. You would have got a very wrong answer. Can you see why?

The reason is that when the bullet enters the sand the force on it is dissipative-friction. There is a big energy transfer which raises the temperature of the bullet and the sand, i.e. a loss of mechanical energy. You might like to work out what proportion of the bullet's original kinetic energy is so transferred. This is a good example of the 'safety' of the momentum principle, but the need for caution in using the energy method.

A Collision

The classic collision between two hard balls is not easy to study in detail, although the general characteristics are well enough known. A detailed study [3] is complicated by the elastic deflection of the spheres while in contact; the analysis by Hertz (1881) is the foundation of the theory of contact stresses. But we can study a 'model' of the situation by once again separating the elasticity from the mass of each body, as in Fig. 6.8.

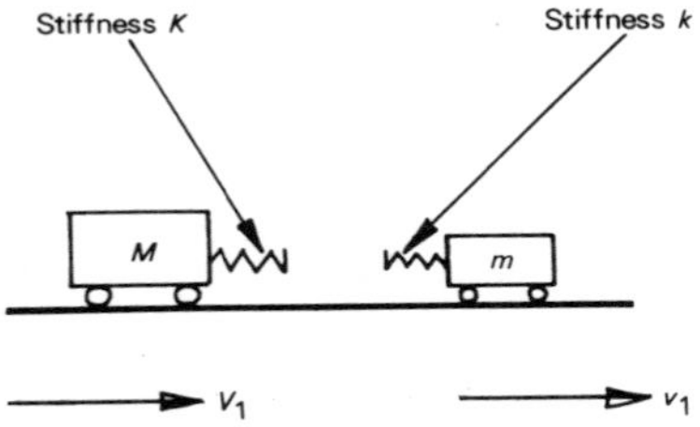

FIG. 6. 8

For our model collision, imagine two rigid railway trucks, masses M and m, fitted with *massless* buffer springs of total stiffness K on truck M and k on truck m (stiffness is force per unit displacement). The trucks are moving along a straight track with velocities V_1 and v_2 before they collide ($V_1 > v_1$). Our model is perfect in that there are no air-resistance of friction forces acting.

Consider first the extreme cases (a) coupling and (b) a perfectly elastic rebound, and finally (c) a rebound with some energy dissipation in the springs.

(a) If the trucks *couple together* on impact there is no problem; their coupled velocity V can be found by conservation of momentum

$$MV_1 + mv_1 = (M + m)V$$

Because there are no external forces acting on the system consisting of both trucks, this expression gives the momentum *at all times* before, during and after the collision.

$$V = \frac{MV_1 + mv_1}{M + m} \tag{6.9}$$

The loss of kinetic energy due to coupling is

$$\tfrac{1}{2}\left[MV_1^2 + mv_1^2 - \frac{(MV_1 + mv_1)^2}{M + m}\right]$$

$$= \tfrac{1}{2}\,\frac{mM}{M + m}\,(V_1 - V_1)^2 \qquad (6.10)$$

So far we have not employed the ideal properties of our model. These momentum considerations apply to any coupling collision. If you have any plasticine left from the earlier experiment you might try it with two plasticine balls.

(b) For a *perfectly elastic rebound* we enlist the properties of our perfect model. We assume that the energy stored in the springs during the collision is fully restored to the trucks as kinetic energy during the rebound. This again is very straightforward; the two expressions given by conservation of momentum, and equality of kinetic energy before and after impact, give the two velocities after impact:

momentum $\quad MV_1 + mv_1 = MV_2 + mv_2$

no energy loss $\quad MV_1^2 + mv_1^2 = MV_2^2 + mv_2^2$

Rewriting $\quad M(V_2 - V_1) = -\,m(v_2 - v_1)$

$$M(V_2^2 - V_1^2) = -\,m(v_2^2 - v_1^2)$$

Dividing $\quad \dfrac{V_2^2 - V_1^2}{V_2 - V_1} = \dfrac{v_2^2 - v_1^2}{v_2 - v_1}$

i.e. $\quad V_2 + V_1 = v_2 + v_1$

so $\quad V_2 - v_2 = -\,(V_1 - v_1)$

In words, the velocity of recession equals the velocity of approach, or the *relative* velocity of the trucks is simply reversed by the collision. Note also, that as a consequence of momentum conservation, the velocity of the combined centre of mass of the two trucks remains unaltered throughout.

It can easily be shown that the final velocities are

$$V_2 = \frac{2mv_1 + (M - m)V_1}{M + m}$$

$$v_2 = \frac{2MV_1 - (M - m)v_1}{M + m} \qquad (6.11)$$

At the point of closest approach of the two trucks, when the springs have their maximum compression, the two trucks will have, for that instant only, a common velocity V,

$$V = \frac{MV_1 + mv_1}{M + m}$$

The decrease of kinetic energy at that time we know from (6.10) to be

$$\frac{1}{2}\frac{mM}{M + m}(V_1 - v_1)^2$$

This decrease is stored in the springs as strain energy. So we can calculate the maximum force in the springs.

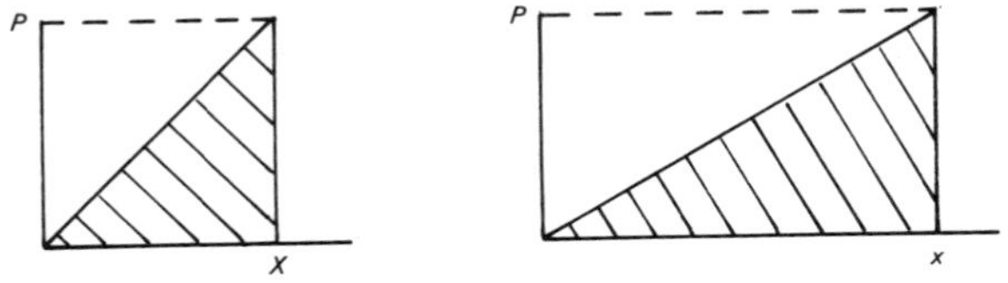

FIG. 6. 9 The buffer springs

If the load/displacement curves of the springs are as shown in Fig. 6.9 the strain energies in the two sets of springs at maximum force are

$$\tfrac{1}{2}PX \text{ and } \tfrac{1}{2}Px$$

where $P = KX = kx$

$$\tfrac{1}{2}PX = \frac{P^2}{2K}, \qquad \tfrac{1}{2}Px = \frac{P^2}{2k}$$

$$\text{Total strain energy} = \frac{P^2}{2}\left(\frac{1}{K} + \frac{1}{k}\right)$$

So $$P^2\left(\frac{1}{K} + \frac{1}{k}\right) = \frac{mM}{M + m}(V_1 - v_1)^2$$

The maximum force is proportional to the initial relative velocity.

If we concern ourselves only with velocities, not forces, we can draw some conclusions about balls from this analysis, provided we are prepared to confine ourselves to perfectly elastic balls. For example, a perfectly elastic golf club striking with velocity V_1 a perfectly elastic golf ball at rest, will give the ball a velocity v_2 given by (6.11)

$$v_2 = \frac{2MV_1}{M + m}$$

If the mass of the striker M is made very much greater than that of the ball m, then $v_2 \to 2V_1$. Plainly if $M >> m$, the velocity of M will be almost unaffected by the collision, and because the relative velocity is reversed, v_2 inevitably tends to $2V_1$.

Billiard balls are dynamically more interesting than golf balls, partly because there are more of them involved and partly because it is easier to see what is going on. Leaving aside the complications connected with the balls rolling, we are concerned now with a situation in which $M = m$. So if one ball strikes ('full-on') another which is at rest, we have $v_1 = 0$ and

$$V_2 = 0, v_2 = V_1 \qquad \text{from (6.11)}$$

The velocities are simply exchanged, leaving the initially moving ball at rest (the cue ball must have no 'top' or 'bottom').

Now imagine a number of balls set up in a straight line as in Fig. 6.10. All the balls except number 1 are initally at rest. It is obvious what will

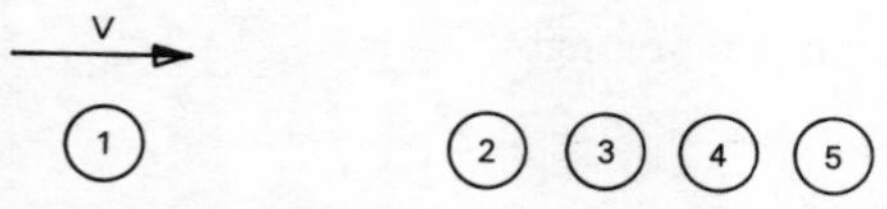

FIG. 6.10 Billiard balls

happen if all the collisions are 'perfect'. The momentum of ball 1 will be exchanged ball-by-ball along the line until number 5 departs with velocity V, leaving the others stationary. The result will be just the same if balls 2 to 5 are all touching initially.

The experiment becomes really interesting with a set-up like that shown in Fig. 6.11. It is difficult to do on a billiard table, but easy on an old baga-

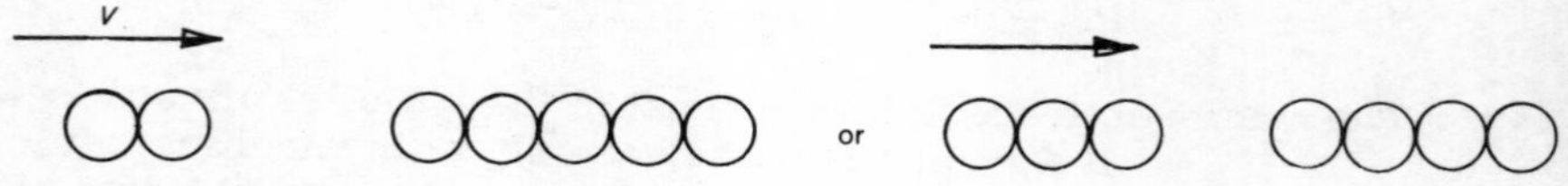

FIG. 6.11 Balls

telle board. It can be done roughly with coins on a smooth table, such as a shove-halfpenny board. But best of all, build yourself, or borrow, a Newton's cradle, a beautiful and enjoyable device, Fig. 6.12.

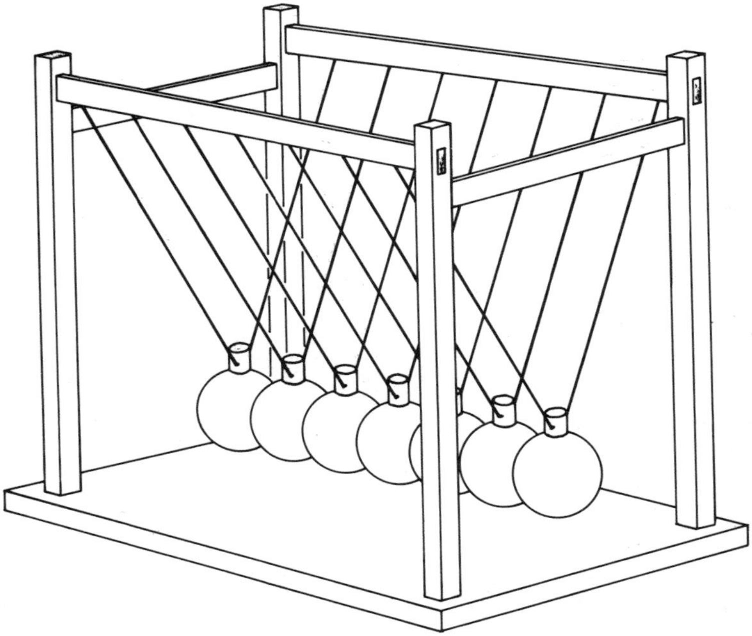

FIG. 6. 12 Newton's cradle

Whichever apparatus you use for these experiments, convince yourself that what happens is in line with the principles in this chapter.

We must return now to our simple model of rigid trucks and massless springs, to examine the last of our cases,

(c) *A rebound with some energy dissipation.* Suppose the springs are such that only a fraction η of the strain energy stored is recovered in the rebound; $\eta = 1$ would apply to perfectly elastic springs, $\eta \approx 0$ to plasticine springs. The progress of the collision would be:

during the approach the kinetic energy would decrease by

$$\frac{1}{2}\frac{Mm}{M+m} \quad (V_1 - v_1)^2$$

and the two trucks would reach a common velocity

$$V = \frac{MV_1 + mv_1}{M + m}$$

during the rebound the kinetic energy would increase by

$$\eta \times \frac{1}{2}\frac{mM}{M + m}(V_1 - v_1)^2$$

and the trucks would separate with final velocities V_2 and v_2. Our job now is to find V_2 and v_2.

It helps to write the expressions for kinetic energy in a form suggested by (6.10).

$$\text{Kinetic energy before collision} = \tfrac{1}{2}\left[MV_1^2 + mv_1^2\right]$$

$$= \frac{1}{2}\frac{(MV_1 + mv_1)^2}{M + m} + \frac{1}{2}\frac{mM}{M + m}(V_1 - v_1)^2$$

$$\text{Similarly, k.e. after collision} = \tfrac{1}{2}\left[MV_2^2 + mv_2^2\right]$$

$$= \frac{1}{2}\frac{(MV_2 + mv_2)^2}{M + m} + \frac{1}{2}\frac{mM}{M + m}(V_2 - v_2)^2$$

It is obvious from momentum conservation that the first terms on the right-hand side of these two expressions are equal, and so the energy change must occur solely in the second terms. Now the second term in the k.e. before collision is the energy which is transferred to the springs as strain energy, and the second term in the k.e. after collision is therefore the transfer from strain energy to kinetic energy in the rebound. It is the ratio of these that we are calling η.

$$\frac{1}{2}\frac{mM}{M + m}(V_2 - v_2)^2 = \eta \times \frac{1}{2}\frac{mM}{M + m}(V_1 - v_1)^2$$

i.e. $$(V_2 - v_2)^2 = \eta(V_1 - v_1)^2$$

Plainly $(V_2 - v_2)$ and $(V_1 - v_1)$ must be of opposite sign, so we want the negative square root of η.

$$(V_2 - v_2) = -\eta^{1/2}(V_1 - v_1) \tag{6.12}$$

This expression and the momentum equation can be solved straightforwardly for V_2 and v_2. In our simple model, η will be less than 1 only because of elastic hysteresis in the springs. For steel springs η would be very close to 1, and $\eta^{1/2}$ would be even closer. But in a real version of our ideal model, the springs would have mass and the trucks would not be rigid, so both would be left vibrating after the collision. The same would be true of real balls. So in a real situation η would be smaller than a single hysteresis loop would suggest. However, for hard bodies, particularly the stiffer metals, η would still be close to unity.

The expression (6.12) is of course the same as the familiar

normal velocity of recession $= e \times$ normal velocity of approach

where e is called the *coefficient of restitution.*

In the *Principia,* Newton describes experiments with balls of steel, cork, glass and 'balls of wool, made up tightly, and strongly compressed'. '... the balls always receding one from the other with a relative velocity, which was to the relative velocity with which they met as about 5 to 9 (wool). Balls of steel returned with almost the same velocity; those of cork with a velocity something less; but in balls of glass the proportion was as about 15 to 16'.

Satellite orbits

In turning away from straight-line motion, we come to the most interesting particle motion, that of a satellite about its mother sun or planet.

This brings us back to the origins of dynamics, because it was Kepler's empirical laws, based on his own and Tycho Brahe's observations of the motion of planets (particularly his own on the orbit of Mars in the first decade of the seventeenth century), that led to the formulation of Newton's laws of motion and gravitation—the reverse therefore of the exercise that we are about to embark on.

The real celestial problem concerns the motion of bodies under the gravitational influence of many other bodies. A simple version is that of two bodies which are sufficiently isolated to be *equally* affected by all other bodies (the earth and moon are an approximation to this); the acceleration of each body relative to the centre of mass of the pair is just that due to the gravitational effect of the other. A still further simplification arises when the mass of one of the isolated pair is very much greater than that of the other. Then the centre of mass of the pair is practically coincident with that of the more massive body, and the motion of the light body relative to the heavy one can be analysed as if it were about a fixed centre in an absolute reference frame; a man-made satellite in orbit around the earth or moon for example.

To find the shape of the orbit, consider a particle of mass m moving in a plane which contains a much greater mass M, and use the centre of mass of M as the origin of polar coordinates, as in Fig. 6.13 (it is implicit in this simple analysis that M is either a particle or a sphere with radial symmetry, as encountered earlier in connection with gravitation).

Equations (5.1) and (5.2) give the radial and transverse components of velocity and acceleration and they are shown on Fig. 6.13 (b) and (c). If $\mathbf{F}$ is

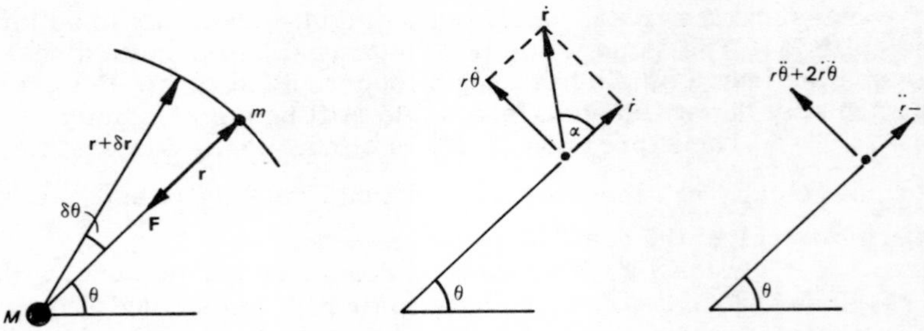

FIG. 6.13

(a) Coordinates and force (b) Velocity (c) Acceleration

the only force on m, then Newton's Second Law applied to the tangential and radial components gives

$$0 = m(r\ddot{\theta} + 2\dot{r}\dot{\theta})$$
$$-\mathbf{F} = m(\ddot{r} - r\dot{\theta}^2)$$

The first gives

$$r\ddot{\theta} + 2\dot{r}\dot{\theta} = \frac{1}{r}\frac{\mathrm{d}}{\mathrm{d}t}(r^2\dot{\theta}) = 0$$

so $$r^2\dot{\theta} = \text{constant} = h, \text{ say} \tag{6.13}$$

$mr^2\dot{\theta}$ is the magnitude of the moment of the momentum of m about the origin. The momentum of m is $m\dot{\mathbf{r}}$ and its moment about the origin is $\mathbf{r} \wedge m\dot{\mathbf{r}}$. The magnitude of this moment of momentum is $|\mathbf{r}||m\dot{\mathbf{r}}| \sin \alpha$ and its value is plain to see in Fig. 6.13 (b); it is $mr^2\dot{\theta}$. So h is the moment of momentum of m divided by m, and it is evidently constant in magnitude for *any* central force.

The area swept out by the radius vector as m moves from $\mathbf{r}$ to $\mathbf{r} + \delta\mathbf{r}$ is

$$\delta A = \tfrac{1}{2} r(r + \delta r) \sin \delta\theta \quad (\text{area of } \triangle ABC = \tfrac{1}{2} ab \sin \hat{C})$$
$$= \tfrac{1}{2} r^2 \delta\theta \quad \text{to first order}$$

The *rate* at which area is swept out, therefore,

$$\frac{\mathrm{d}A}{\mathrm{d}t} = \frac{1}{2} r^2 \frac{\mathrm{d}\theta}{\mathrm{d}t} = \frac{h}{2} = \textit{constant}$$

So under any central force, the line joining m to M sweeps out equal areas in equal times. This was one of Kepler's observations for the sun's planets.

For the particular case of the inverse square law of gravitation, the orbit shape can be found by solving the differential equation

$$m(\ddot{r} - r\dot{\theta}^2) = -F = -\frac{GMm}{r^2}$$

or

$$\ddot{r} - r\dot{\theta}^2 = -\frac{GM}{r^2} = -\frac{\mu}{r^2}, \text{ say,} \tag{6.14}$$

The standard solution involves getting rid of one of the variables, t, and changing one of the others, r, to $u = 1/r$.

From (6.13) $\dot{\theta} = \frac{h}{r^2} = hu^2$

$$\dot{r} = \frac{dr}{dt} = \frac{dr}{d\theta} \cdot \frac{d\theta}{dt} = \dot{\theta}\frac{dr}{d\theta} = \frac{h}{r^2}\frac{dr}{d\theta} = -h\frac{d}{d\theta}\left(\frac{1}{r}\right)$$

$$\ddot{r} = \frac{d}{dt}(\dot{r}) = \frac{d\theta}{dt} \cdot \frac{d(\dot{r})}{d\theta} = -\frac{h^2}{r^2}\frac{d^2}{d\theta^2}\left(\frac{1}{r}\right) = -h^2u^2\frac{d^2u}{d\theta^2}$$

(6.14) becomes, on substitution,

$$\frac{d^2u}{d\theta^2} + u = \frac{\mu}{h^2} \text{ (a constant)}$$

This is an ordinary second order differential equation. Even for those with no experience of solving differential equations, the solution of this one is easy to see if it is rewritten

$$\frac{d^2}{d\theta^2}\left(u - \frac{\mu}{h^2}\right) = -\left(u - \frac{\mu}{h^2}\right)$$

which is an example of

$$\frac{d^2y}{d\theta^2} = -y$$

The solution of this is

$$y = A\cos\theta + B\sin\theta$$

or

$$y = A\cos(\theta + \phi)$$

where A and ϕ are arbitrary constants. So the solution we require is

$$u - \frac{\mu}{h^2} = A\cos(\theta + \phi)$$

The line $\theta = 0$ can be selected for convenience, so let it be such that $\phi = 0$. Reverting now to the variable r,

$$\frac{h^2}{\mu r} = 1 + \frac{Ah^2}{\mu} \cos\theta \tag{6.15}$$

Equation (6.15) is the general polar equation for conic sections, written in a way which may not be familiar to all engineering undergraduates.

Conic Sections

Any of the conic sections can be formed by the locus of a point which moves in such a way that its distance from a fixed point (the *focus*) is in a constant ratio to its distance from a fixed line (the *directrix*). In Fig. 6.14, S is the focus and MD is the directrix. The locus of P is given by

$$\mathrm{SP} = e\mathrm{PM}$$

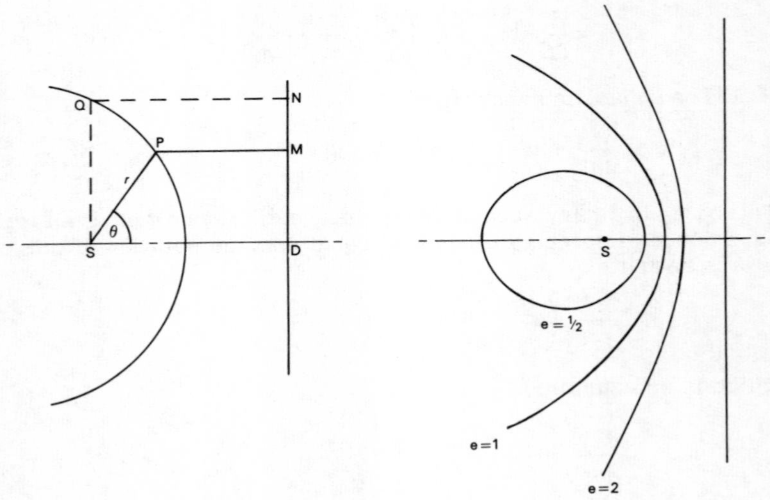

FIG. 6.14 Conic sections

The ratio e is called the *eccentricity*.

If $e > 1$ the curve is one branch of a hyperbola,
if $e = 1$ the curve is a parabola,
if $1 > e > 0$ the curve is an ellipse,
and if $e = 0$ the curve is a circle.

The first two are open curves, extending to infinity, the last two are closed curves. The parabola can be thought of as the boundary between a family of ellipses on one side and a family of hyperbolae on the other. The polar equation with S as origin is easily derived.

$$r = \mathrm{SP} = e\mathrm{PM}$$
$$\mathrm{PM} = \mathrm{SD} - r\cos\theta = \mathrm{QN} - r\cos\theta$$
$$e\mathrm{QN} = \mathrm{SQ} = l\text{, the semi-latus rectum}$$
$$r = l - er\cos\theta$$

Or
$$\frac{l}{r} = 1 + e\cos\theta \tag{6.16}$$

Equations (6.15) and (6.16) are the same therefore, with $l = h^2/\mu$ and $e = Ah^2/\mu$.

Another of Kepler's observations was that each of the planets moves in an elliptical orbit with the sun at one focus.

Now let us remind ourselves what the various terms mean. r and θ are clear enough. h is the (constant) moment of momentum about the origin divided by the mass, $\mu = GM$ and is a true constant for a given M, and A is a constant of integration whose value will depend on the 'initial conditions' of the motion (in fact the conditions at any specified time).

To determine which of the shapes the orbit takes, consider the energy of the particle. If the particle is flying clear of any atmosphere, there will be no dissipative forces acting and the energy equation can be used in the form k.e. + p.e. = constant.

$$T + V = mE,\ \text{a constant}$$

With a datum at infinity, we have a value for the potential energy in equation (6.7)

$$V = -\frac{GMm}{r} = -\frac{\mu m}{r}$$

The kinetic energy is

$$T = \tfrac{1}{2}\,mv^2 = \tfrac{1}{2}\,m(r^2\dot{\theta}^2 + \dot{r}^2)$$

so
$$E = \frac{1}{2}\,(r^2\dot{\theta}^2 + \dot{r}^2) - \frac{\mu}{r}$$

We have already seen that

$$\dot{r} = h\,\frac{\mathrm{d}u}{\mathrm{d}\theta}$$

$$\dot{\theta} = hu^2,\ \text{so}\ \ r^2\dot{\theta}^2 = h^2u^2$$

Therefore
$$E = \frac{1}{2}\left\{u^2 + \left(\frac{\mathrm{d}u}{\mathrm{d}\theta}\right)^2\right\}h^2 - \mu u$$

but (6.16) is $u = \frac{1}{l}(1 + e\cos\theta)$

$$\frac{du}{d\theta} = -\frac{e}{l}\sin\theta$$

$$E = \frac{h^2}{2l^2}\left\{1 + 2e\cos\theta + e^2\cos^2\theta + e^2\sin^2\theta\right\} - \frac{\mu}{l}(1 + e\cos\theta)$$

Remembering that $l = \frac{h^2}{\mu}$, this reduces to

$$E = \frac{\mu}{2l}(e^2 - 1) \tag{6.17}$$

For an ellipse, $e < 1$, E is negative,
for a parabola, $e = 1$, E is zero, and
for a hyperbola, $e > 1$, E is positive.

Kinetic energy is necessarily positive, potential energy is negative, tending to zero at infinity, so if the particle is to reach infinity, i.e. 'escape' from M or avoid 'capture' by M, the energy E must be greater than or equal to zero. What this means in terms of velocity we can see by returning to the form

$$mE = \frac{1}{2}mv^2 - \frac{\mu m}{r}$$

which gives

$$v^2 = \frac{2\mu}{r} + 2E$$

Substituting for E from (6.17)

$$v^2 = \frac{2\mu}{r} + \frac{\mu}{l}(e^2 - 1) \tag{6.18}$$

v^2 will be *less than*, *equal to*, or *greater than* $\frac{2\mu}{r}$

according to whether

e is *less than*, *equal to*, or *greater than* unity.

So for a particle with velocity V at radius R, the orbit will be an *ellipse*, a *parabola*, or a *hyperbola*, if V^2 is *less than*, *equal to* or *greater than* $2\mu/R$ respectively.

The minimum escape velocity at radius R is therefore

$$V_e = \sqrt{\frac{2\mu}{R}} \tag{6.19}$$

Elliptical and circular orbits

The closed orbits are elliptical, or in the limit circular. Consider first the *circular orbit.*

(6.16) $$\frac{l}{r} = 1 + e \cos \theta$$

(6.15) $$\frac{h^2}{\mu r} = 1 + \frac{Ah^2}{\mu} \cos \theta$$

(6.17) $$E = \frac{\mu}{2l}(e^2 - 1) = \frac{1}{2}v^2 - \frac{\mu}{r}$$

For a circular orbit, $e = 0$ and $l = r =$ constant.

$$E = -\frac{\mu}{2l} = -\frac{\mu}{2r} = \frac{1}{2}v^2 - \frac{\mu}{r}$$

so

$$v^2 = \frac{\mu}{r}$$

For a circular orbit at radius R, therefore, the velocity must be

$$V = \sqrt{\frac{\mu}{R}}$$

This could have been found much more easily of course as an example of motion in a circle.

Note that this circular orbit v^2 is just half the escape v^2. In other words, to reach escape velocity from a circular orbit, the velocity must be increased by a factor of $\sqrt{2}$.

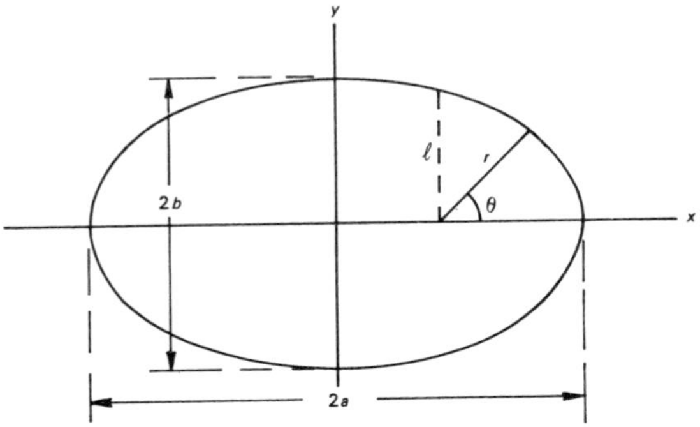

FIG. 6. 15 The ellipse

The general closed orbit is the *ellipse*, with M at one focus. Perhaps the more familiar representation of an ellipse is

$$\frac{x^2}{a^2} + \frac{y^2}{b^2} = 1$$

where $2a$ and $2b$ are the major and minor axes, and the two foci are at $x = \pm ea$. By referring to Fig. 6.15 we can compare the polar and cartesian representations. The length of the major axis, $2a$, is clearly $(r)_{\theta = 0} + (r)_{\theta = \pi}$

$$\frac{l}{r} = 1 + e \cos \theta$$

so

$$2a = \frac{l}{1+e} + \frac{l}{1-e} = \frac{2l}{1-e^2}$$

Therefore $\quad l = a(1 - e^2)$

To find the velocity, substitute for l in equation (6.18)

$$v^2 = 2\mu\left(\frac{1}{r} - \frac{1}{2a}\right) \tag{6.20}$$

The velocity is a maximum when r is a minimum and vice versa. The values are easily found by substitution.

The various velocities can be put in 'earthly' terms by remembering that $\mu = GM$ and that $g = GM/r^2$ (equation (3.3)), being the acceleration due to gravity at any radius r greater than the earth's radius. If we treat the value of g at the earth's surface, and the radius of the earth, as constants, and denote them by g_0 and R_0, then the various velocities can be written:

elliptical orbit $\quad v^2 = 2g_0R_0^2\left(\frac{1}{r} - \frac{1}{2a}\right)$

circular orbit of radius R $\quad V^2 = \frac{g_0R_0^2}{R}$

escape velocity from radius R $\quad V_e^2 = \frac{2g_0R_0^2}{R}$

As a simple exercise, work out some of the figures that are often quoted in the press in connection with space-shots. Use the suffix 0 to apply to the surface of either the earth or the moon.

Streams of particles, jet and rocket engines

In all sorts of engineering situations, we meet streams of particles going into and coming out of solid bodies, or interacting with them in some way.

They are usually fluid, in turbomachinery, jets, rockets and so on. This is not the place to write about fluid mechanics, but we can tackle the dynamics of these devices by simple applications of impulse and momentum.

Consider first a body at rest, issuing a jet of liquid as in Fig. 6.16(a), at velocity v and at a mass rate m' (mass per unit time). An element of fluid which emerges in time δt will have mass $m'\delta t$. Isolating the body and the element

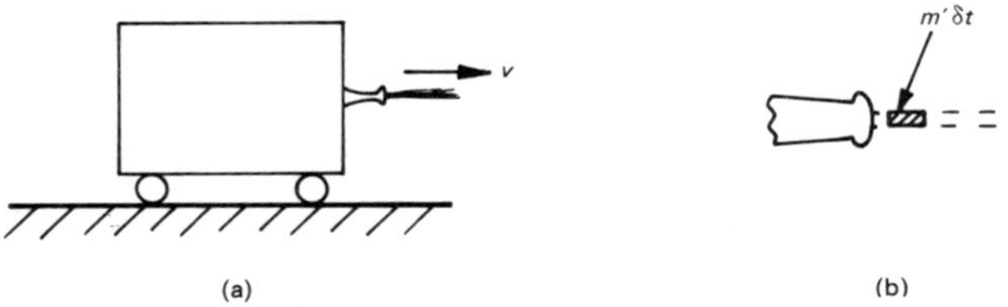

FIG. 6.16

as two particles, the change of momentum of the element is $m'\delta tv$. (Note that this element could just as well be one of a stream of solid particles.) This is the impulse $F\delta t$ over the time δt. By Newton's Third Law the two particles experience equal and opposite impulses, if there are no external forces. If the body is to remain at rest, therefore, there must be a restraining force F. In the absence of a restraining force, the body will be accelerated leftwards in the diagram by a 'thrust'

$$F = m'v$$

The thrust on the body (or 'rate of impulse') is equal to the rate of change of momentum of the jet.

Now apply the same principle to a vane or blade, which changes the direction of a jet of fluid, Fig. 6.17.

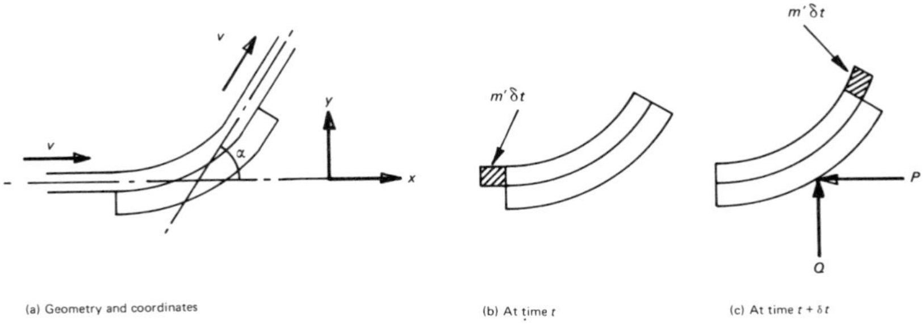

FIG. 6.17

Neglecting friction losses, the jet will leave the vane at the same velocity as at entry, v, but deflected through an angle α. Referring to Fig. 6.17 (b) and (c), the change of momentum of the jet in time δt is:

In the x direction: $m'\delta t\ (v \cos\alpha - v) = F_x \delta t$
In the y direction: $m'\delta t\ v \sin\alpha = F_y \delta t$

The rates of change of momentum, i.e. the 'impulse rates', on the jet are equal and opposite to the force components on the blade. To keep the blade at rest, external force components P and Q must be applied

$$P = -F_x = m'v(1 - \cos\alpha)$$

$$Q = F_y = m'v \sin\alpha$$

How would these expressions be modified if the jet lost 10 per cent of its entry velocity during its contact with the blade; i.e. if it left the blade at $0.9\,v$?

Finally we examine two cases in which the body concerned is in motion, scooping up and/ or ejecting material as it goes, *jet and rocket engines*.

While the details of jet and rocket engines are complicated, the basic dynamics can be examined in a quite simple way. Both achieve a propulsive force by slinging rearwards a continuous stream of gas. They differ however in an important way; both are jet engines, but the so-called jet engine takes in its 'reaction mass' (the mass it slings rearwards) in the form of air as it goes along (it needs an atmosphere therefore), whereas the rocket carries it, continuously generating its gas jet by burning propellant.

The jet engine burns fuel of course, but the mass rate of consumption of fuel is very small compared with the rate of intake of air and output of exhaust, so for our purpose it is in order to neglect it.

Let the mass of the jet engine, plus its share of the mass of the aircraft, be M and let the mass flow rate of gas through it be m' (mass/unit time); Fig. 6.18.

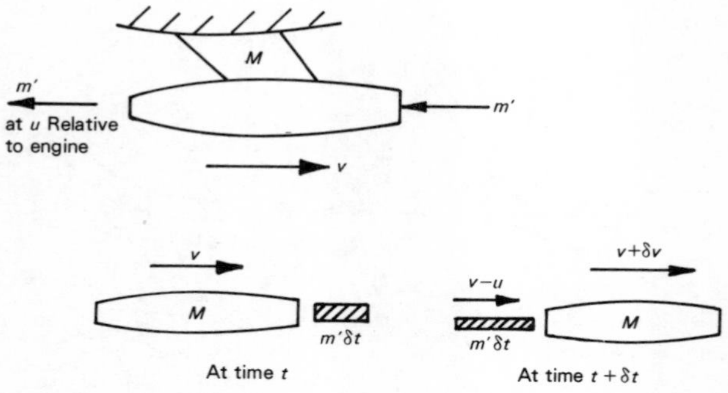

FIG. 6.18 Jet engine

In time δt the engine takes in $m'\delta t$ from *rest* and discharges it at u relative to the engine, rearwards, i.e. with an absolute forward velocity of $(v - u)$. In horizontal flight the only external force in the direction of motion will be air resistance, which we denote by $-R$.

Isolating a fixed mass $M + m'\delta t$ and applying impulse-momentum over time δt,

$$F\delta t = \text{change of momentum in time } \delta t$$

$$-R\delta t = M(v + \delta v) + m'\delta t(v - u) - Mv$$

$$-R = M\frac{\delta v}{\delta t} + m'(v - u)$$

$$M\frac{dv}{dt} = -R + m^1(u - v) \qquad (6.21)$$

The last term is the jet engine thrust, and to be positive, it must have $u > v$.

Finally, we turn to the *rocket engine*. First consider the situation corresponding to the jet engine in the last example: i.e. motion in a straight line with an external force $-R$ in the direction of motion. Now the reaction mass is being discharged from within the rocket, so the loss of mass will not be negligible and must be taken into account; see Fig. 6.19.

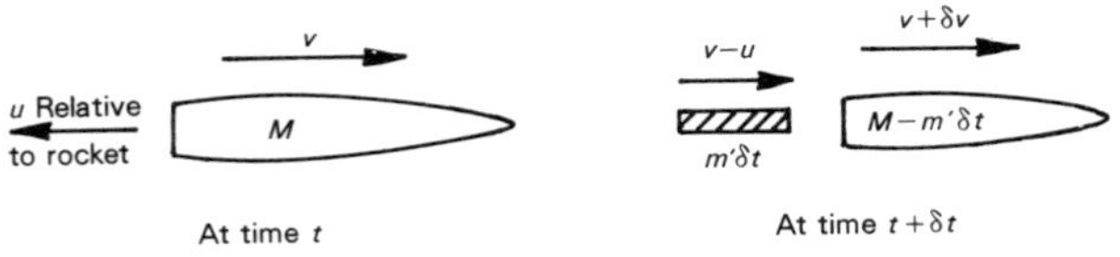

FIG. 6.19 Rocket engine

Isolating again a fixed mass during the time increment δt, impulse = change of momentum

$$-R\delta t = (M - m'\delta t)(v + \delta v) + m'\delta t(v - u) - Mv$$

$$= \cancel{Mv} - m'\cancel{v}\delta t + M\delta v + m'\cancel{v}\delta t - m'u\delta t - \cancel{Mv}$$

$$+ \text{ second order term}$$

$$= M\delta v - m'u\delta t$$

which gives

$$M\frac{dv}{dt} = -R + m'u \qquad (6.22)$$

Note that the rocket engine is not subject to the limitation that the jet exhaust velocity must be greater than the rocket velocity; compare equations (6.21) and (6.22). In addition to this difference between (6.21) and (6.22), remember the other important difference that M is constant in (6.21), but varying in (6.22). To illustrate this, consider the simplest example of (6.22), viz. $R = 0$.

$$M\frac{dv}{dt} = m'u$$

Plainly $m' = -\,dM/dt$, the rate of loss of material

$$\frac{dv}{dt} = -\frac{u}{M}\frac{dM}{dt}$$

$$dv = -u\frac{dM}{M}$$

Integrating between times 1 and 2

$$\int_{v_1}^{v_2} dv = -u\int_{M_1}^{M_2}\frac{dM}{M}$$

$$v_2 - v_1 = -u\ln\frac{M_2}{M_1} = u\ln\frac{M_1}{M_2} \tag{6.23}$$

We have assumed here that u is constant, but *not* that m' (i.e. dM/dt) is constant.

Think about the advantage, in terms of attainable velocity, to be gained from using a multi-stage rocket. Can you produce expressions corresponding to (6.23) for two- and three-stage rockets? Assume that the empty body of a burnt out stage simply parts company, without a kick, as the subsequent stage ignites.

Think also about the effect of gravity in a vertical ascent. Do this by going back to (6.22) and putting $R = Mg$ and neglecting air resistance; then integrate.

Bibliography

1. Hoyle, R. D. and Clarke, P. H., *Thermodynamic Cycles and Processes*, Longmans, 1972
2. Inglis, C., *Applied Mechanics for Engineers*, Cambridge University Press, 1951, Dover, 1963
3. Timoshenko, S. and Goodier, J. N., *Theory of Elasticity*, McGraw-Hill, 1951
4. Jeans, J. H., *Theoretical Mechanics*, Ginn, 1907
5. Rutherford, D. E., *Classical Mechanics*, Oliver and Boyd, 1951

Examples

The acceleration due to gravity near the earth's surface can be taken as $9.81\ \mathrm{m/s^2}$.

Ex. 6.1 The 350 kg hammer of a pile-driving rig falls 2 m from rest on to the top of a 150 kg pile, and moves with the pile without rebound. Estimate the velocity of the hammer and pile immediately after impact. [4.4 m/s]

Ex. 6.2 At a fairground, children can slide on a mat down a helical slide after climbing a lot of steps up the middle (a helter-skelter).

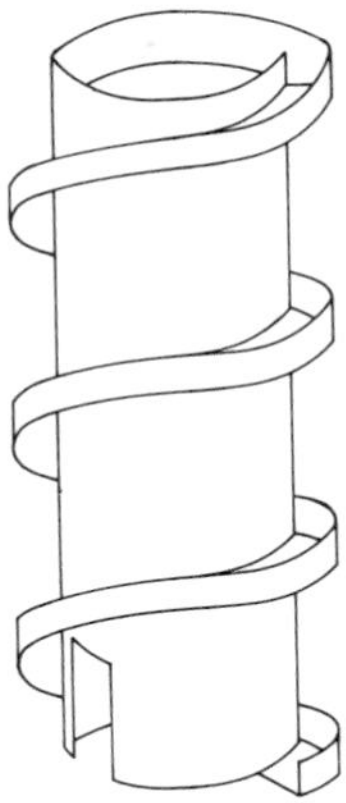

Ex. 6. 2

Consider a simple cylindrical helter-skelter in which the height is 6 m, there are three complete turns of the track round the cylinder, and the radius to the centre of the track is 1.5 m.

Calculate the velocity of a child reaching the bottom (a) if there is no friction, (b) if the coefficient of friction between mat and track is 0.1, and (c) if the coefficient is 0.2. [10.9, 7.7 m/s]

Ex. 6. 3 A railway truck of mass 5,000 kg moving at 4 m/s strikes a second truck of mass 6,000 kg, initially at rest. Both trucks have two buffer springs at each end, and all the springs have a stiffness of 3×10^6 N/m. Calculate the velocities of the trucks after separation, and the maximum force between the trucks during the impact, assuming no energy loss.

[− 0.36 m/s, + 3.64 m/s, 3.6×10^5 N]

Ex. 6. 4 Derive an expression for the periodic time of a satellite in an elliptical orbit, in terms of the major axis of the ellipse.

[$T = 2\pi a^{3/2}/(GM)^{1/2} = 2\pi a^{3/2}/R_0\sqrt{g_0}$]

Ex. 6.5 An aircraft jet engine has a mass airflow of 100 kg/s and a jet stream velocity of 600 m/s. Plot the theoretical thrust of the engine against aircraft speed. How do you think this might be modified in reality?

Ex. 6.6 A rocket of mass 5,000 kg takes off vertically from rest. Of the total mass 4,000 kg is propellant, and this burns at a uniform rate in 100 s. The jet exhaust velocity is 2,000 m/s. Neglecting air resistance, estimate the acceleration just before all-burnt, and the velocity and altitude at all-burnt.

[70 m/s^2, 2 240 m/s, 70 km]

Ex. 6.7 Derive expressions for the thrust on the stationary deflectors in the diagram, in terms of the jet velocity and flow rate. Neglect losses.

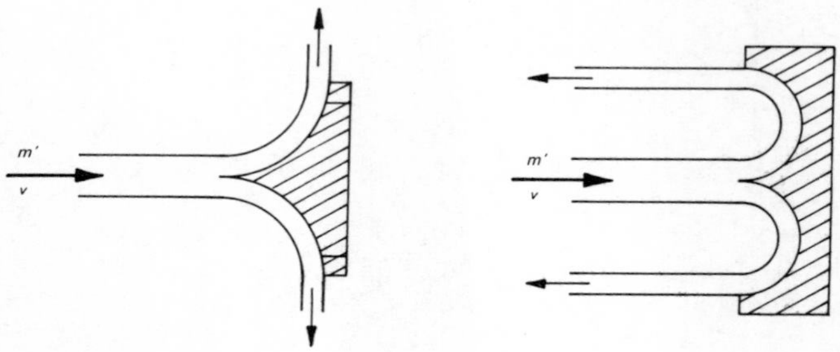

Ex. 6.7

Ex. 6.8 Coal falls vertically on to a horizontal conveyor belt at a rate of 10 kg/s. Apart from friction forces, what force would be needed to keep the conveyor belt moving at 1 m/s? What would the force be if there were a rise of 1 m over the length of the belt? [10 N, 108 N]

7 Body Dynamics

The Newtonian dynamics of the *Principia* was almost entirely concerned with particles. During the eighteenth century the subject was developed to cover extended bodies; prominent amongst the long list of distinguished contributors to the development were Euler, d'Alembert and Lagrange.

When we look at body dynamics in engineering, we find that in modern practice, particularly in high speed machinery, the majority of moving parts have an inherent simplicity of shape; partly for design reasons (largely dynamic) and partly because most of the methods of manufacture of the parts produce flat surfaces and solids of revolution. Furthermore, most of the bodies which rotate do so about fixed axes.

So the majority of engineering dynamics situations can be analysed in terms of quite simple two-dimensional theory. The layout of this chapter will therefore be as follows: a number of simple cases, which cover a large proportion of real situations, will be examined using theoretical results which will sometimes be given simply as statements, without proof; followed by a more general three-dimensional approach, which will serve the dual purpose of proving the earlier statements and providing the methods to tackle one or two of the exceptional, more complicated, cases; finally the limitations of the assumption of rigidity in dynamics will be briefly examined.

Plane Motion

Under this heading come the motions of bodies in which all points in the body move in parallel planes; any line in the body originally at right angles to those planes remains at right angles to them throughout the motion. The motion can be completely described in a single reference plane, the plane in which the centre of mass moves. The applied forces driving the motion have their resultant in the reference plane, although in some cases we shall see that couples out of the plane are required to constrain the motion to the plane.

All the externally applied forces will have a resultant in the reference plane, and this can be replaced by a parallel force F acting through the centre of mass plus a couple M with axis perpendicular to the plane. Newton's second law extended from particles to solid bodies gives the result

$$\begin{aligned} \mathbf{F} &= m\mathbf{a} \\ M &= I\ddot{\theta} \end{aligned} \tag{7.1}$$

$\mathbf{a}$ is the acceleration of the centre of mass, $\ddot{\theta}$ the angular acceleration. M and $\ddot{\theta}$ need not be shown as vectors since their direction can only be perpendicular to the reference plane. I is the *moment of inertia* of the body measured about an axis through its centre of mass and perpendicular to the plane; there will be more about this quantity below for those who are not familiar with the term.

Before pursuing general plane motion, it will be wise to look at the two elements of the motion separately, viz. translation only (no rotation), and rotation only (about a fixed axis).

d'Alembert's Principle

The way ahead will be made smoother by the introduction at this stage of *d'Alembert's Principle*. d'Alembert's own statement of the principle was abstruse, but in essence it is the rewriting of Newton's Second Law for a particle in the form

$$\mathbf{F} + (-m\mathbf{a}) = 0$$

This looks trivial, but implies that the equation of motion of the particle can be written as an equation of 'equilibrium' if the acceleration term is embodied in the equation in the form $(-m\mathbf{a})$, and treated as a fictitious 'force'. Of course it is not a force, but it has the same dimensions as a force and is often called an 'inertia force' or a 'd'Alembert force'. d'Alembert's principle becomes most helpful, as we shall see, when it is extended to the summation of particles that constitute a solid body. Then equation (7.1) can be rewritten

$$\begin{aligned} \mathbf{F} + (-m\mathbf{a}) &= 0 \\ M + (-I\ddot{\theta}) &= 0 \end{aligned} \tag{7.2}$$

Then on a free-body diagram the fictitious d'Alembert force $(-m\mathbf{a})$ and couple $(-I\ddot{\theta})$ can be inserted, and the equations of motion written by asserting 'the sum of all the forces, including the d'Alembert force, and the sum of all the moments of the forces about an arbitrary point, including those of the d'Alembert force and couple, are zero'.

Translation

Translational motion in a plane is by definition motion in which $\dot{\theta} = \ddot{\theta} = 0$. This requires that $M = 0$ and so the equation of motion is

$$\mathbf{F} = m\mathbf{a}$$

with the additional condition that the resultant force **F** must pass through the centre of mass to ensure that $M = 0$. In cartesian coordinates with the xy plane as the reference plane

$$F_x = m\ddot{x}$$
$$F_y = m\ddot{y}$$
$$M_C = 0 \tag{7.3}$$

One of the few motions of interest under this heading is that of a vehicle on a straight road or rail track. For the present we will neglect the effects of the rotating parts, all internal frictions at bearings, and air resistance.

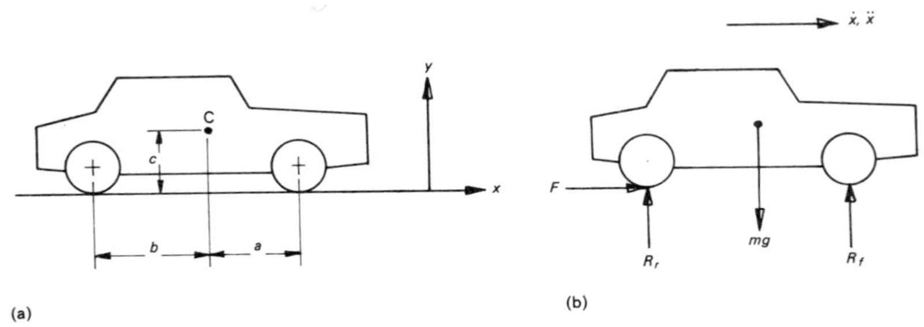

FIG. 7.1 Motor car

Figure 7.1 (a) shows the relevant dimensions of a motor car; Fig. 7.1 (b) shows the forces acting on the vehicle when it is being accelerated by a rear wheel drive. All the wheel forces are assumed to be equally shared by off-side and near-side. Application of (7.3) gives

$$F = m\ddot{x}$$
$$R_r + R_f - mg = 0$$
$$bR_r - cF - aR_f = 0 \tag{7.4}$$

These can be solved to give the forces at the front and rear wheels in terms of the vehicle acceleration $\ddot{x}$.

$$R_f = \frac{m\,(bg - c\ddot{x})}{a + b}$$

$$R_r = \frac{m\,(ag + c\ddot{x})}{a + b} \tag{7.5}$$

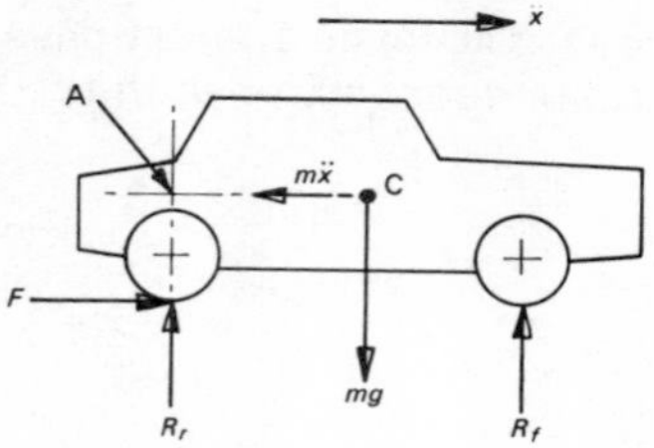

FIG. 7. 2

It is worth noting in passing that even in this very simple case a slight simplification in the solution is achieved by putting the d'Alembert forces on the free-body diagram. In this case there is only $(-m\ddot{x})$ and that is shown in Fig. 7. 2 as a dashed line to indicate that it is not a real force. Of course having taken this step we must be careful not to put in any acceleration terms in the equations *as well*, thereby including them *twice*!

The simplification is achieved in Fig. 7. 2 by taking moments about A, rather than about C, giving the expression for R_f directly

$$F = m\ddot{x}$$

$$R_r + R_f - mg = 0$$

Moments about A,

$$bmg - (a + b)R_f - cF = 0$$

$$R_f = \frac{m\,(bg - c\ddot{x})}{a + b}$$

Returning now to the expressions for R_f and R_r in (7. 5), you will see that changing F from the rear wheels to the front wheels in Fig. 7. 1 (b) would make no difference to any of the expressions. The difference appears when we remember that F has an upper limit of μR_r for rear-wheel drive or μR_f for front-wheel drive, however great the torque from the engine may be; μ is the coefficient of friction between tyre and road. Equations (7. 5) show that the larger the acceleration, the larger the *increase* in R_r and the decrease in R_f.

If instead of accelerating the car we now apply the brakes, the sign of $\ddot{x}$ is changed and R_f increases while R_r decreases, their sum always being equal to mg. It is usual to have brakes on all four wheels. If the coefficient of friction is the same at all the tyre/road contacts, and if the braking system applies the same braking torque to all the wheels, think about which wheels would start slipping first as the braking torque is steadily increased. What parts do the values a, b and c play? Think about racing cars, bicycles and going over the handle-bars.

If the vehicle is going up or down an incline α, equations (7.4) must be modified to

$$F - mg \sin \alpha = m\ddot{x}$$

$$R_r + R_f - mg \cos \alpha = 0$$

$$bR_r - cF - aR_f = 0$$

We turn now to the much more substantial special case of

Rotation about a fixed axis

The following two-dimensional analysis applies strictly only to 'plane bodies', but we shall see later that the results can be applied to real solid bodies. The body to be considered is shown in Fig. 7.3. r, θ, z coordinates will be used, with unit vectors $\hat{\mathbf{r}}, \hat{\boldsymbol{\theta}}, \mathbf{k}$. The body rotates about the z axis. The centre of mass C is not necessarily on the z axis, but it is in the plane $z = 0$.

To examine the motion of the body we imagine it made of a large number of particles, just as we did the body in Fig. 4.2. Furthermore, if we put in the fictitious d'Alembert forces we can proceed in exactly the same way as we did in finding the conditions of equilibrium of the body in Fig. 4.2, but of course now we shall be finding the equation of *motion* of the body in Fig. 7.3.

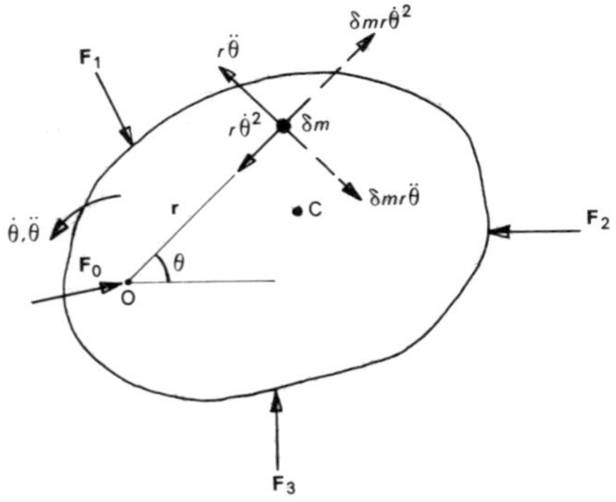

FIG. 7.3

An element of the body δm at $\mathbf{r}$ has acceleration

$$r\dot{\theta}^2(-\hat{\mathbf{r}}) + r\ddot{\theta}(\hat{\boldsymbol{\theta}})$$

The 'd'Alembert forces' are as shown by the dashed lines. The equation of motion of the element is

$$\mathbf{F} + \mathbf{f} + \delta m[r\dot{\theta}^2(\hat{\mathbf{r}}) - r\ddot{\theta}(\hat{\boldsymbol{\theta}})] = 0 \qquad (7.6)$$

where $\mathbf{F}$ is the resultant external force and $\mathbf{f}$ the resultant internal force on the element as we met in Fig. 4. 2. We want to know the moment about the axis, and the force at the axis, to maintain the motion $\dot{\theta}$, $\ddot{\theta}$ of the body. Consider first the *moment*. Taking moments about the z axis of the terms in (7. 6),

$$\mathbf{r} \wedge \mathbf{F} + \mathbf{r} \wedge \mathbf{f} = \delta m \mathbf{r} \wedge r\ddot{\theta}(\hat{\boldsymbol{\theta}})$$

For the whole body therefore

$$\Sigma \mathbf{r} \wedge \mathbf{F} + \Sigma \mathbf{r} \wedge \mathbf{f} = \Sigma \delta m \mathbf{r} \wedge r\ddot{\theta}(\hat{\boldsymbol{\theta}})$$

Over the whole body, all the internal forces (the $\mathbf{f}$'s) will cancel in the summation, and so therefore will their moments. Also, $\mathbf{r} \wedge \hat{\boldsymbol{\theta}} = r(\mathbf{k})$, of magnitude r and perpendicular to both $\mathbf{r}$ and $\hat{\theta}$, so the moment equation becomes

$$\Sigma \mathbf{r} \wedge \mathbf{F} = \mathbf{M} = \Sigma \delta m r^2 \ddot{\theta}(\mathbf{k})$$

where $\mathbf{M}$ is the total moment about the z axis of all the external forces. Since the line of both vectors is the z axis, the equation can be written

$$M_z = \ddot{\theta} \Sigma r^2 \delta m$$

For a continuous rigid body, $\Sigma r^2 \delta m = \int r^2 \mathrm{d}m$ and is called the *moment of inertia* about the z axis, and is denoted by I_z.

So $$M_z = I_z \ddot{\theta} \qquad (7.7)$$

This can be thought of as the rotational equivalent of Newton's Second Law.

Moment of momentum

The velocity of the element δm is $r\dot{\theta}(\hat{\theta})$ and so its momentum is $\delta m r\dot{\theta}(\hat{\boldsymbol{\theta}})$. The moment of the momentum about O is

$$\mathbf{r} \wedge \delta m r\dot{\theta}(\hat{\theta}) = \delta m r^2 \dot{\theta}(\mathbf{k})$$

The total moment of momentum for the body is

$$\mathbf{H} = \dot{\theta}\left(\int r^2 \mathrm{d}m\right)(\mathbf{k})$$

or $$H_z = I_z \dot{\theta}$$

So $$M_z = \dot{H}_z \qquad (7.8)$$

Plainly $$\int_1^2 M_z \mathrm{d}t = H_{z_2} - H_{z_1} \qquad (7.9)$$

This is the time integral of the moment. The rotation integral is given in the next paragraph.

Kinetic energy

The kinetic energy of the element is $\tfrac{1}{2}\delta m r^2 \dot{\theta}^2$. The total kinetic energy of the body is

$$T = \tfrac{1}{2}\dot{\theta}^2 \int r^2 \mathrm{d}m = \tfrac{1}{2} I_z \dot{\theta}^2$$

Now $$M_z = I_z\ddot{\theta} = I_z\dot{\theta}\frac{d\dot{\theta}}{d\theta} = \frac{d}{d\theta}(^1/_2 I_z\dot{\theta}^2)$$

So $$\int_1^2 M_z d\theta = T_2 - T_1 \qquad (7.10)$$

Returning now to the force at the axis of rotation, the sum of the forces on the body can be written as

$$\mathbf{F_0} + \Sigma\mathbf{F}_{1,2}, \text{ etc.}$$

where $\mathbf{F_1}$, etc., are the applied forces, and $\mathbf{F_0}$ is the force which must be provided by the bearings to maintain the position of the axis of rotation. Summing the equation of motion for an element (7. 6) over the whole body, and noting that $\Sigma\mathbf{f} = 0$,

$$\mathbf{F_0} + \Sigma\mathbf{F}_{1,2}, \text{ etc.} + \Sigma\delta m[r\dot{\theta}^2(\hat{\mathbf{r}}) - r\ddot{\theta}(\hat{\theta})] = 0$$

The force at the axis is therefore

$$\mathbf{F_0} = -\Sigma\mathbf{F}_{1,2}, \text{ etc.} - \dot{\theta}^2\int r(\hat{\mathbf{r}})dm + \ddot{\theta}\int r(\hat{\theta})dm$$

The first term on the right-hand side is clear enough and needs no further comment. The last two terms can be rewritten

$$-\dot{\theta}^2\int \mathbf{r}dm + \ddot{\theta}\int \mathbf{k}\wedge\mathbf{r}dm$$

They can be evaluated by replacing $\mathbf{r}$ by $\mathbf{s} + \boldsymbol{\rho}$, which is illustrated in Fig. 7. 4. C is the centre of mass.

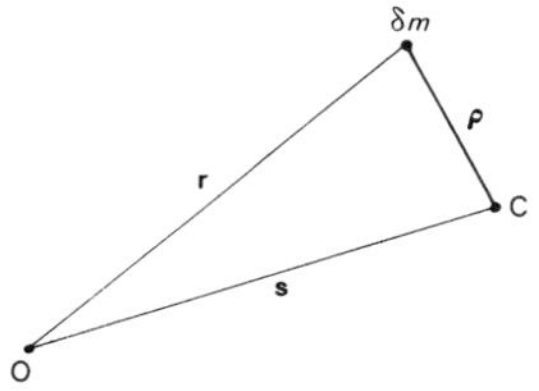

FIG. 7. 4

$$\mathbf{r} = \mathbf{s} + \boldsymbol{\rho}$$

$$\int \mathbf{r}dm = \int \mathbf{s}dm + \int \boldsymbol{\rho}dm = \mathbf{s}\int dm + 0 = \mathbf{s}m$$

because $\mathbf{s}$ is the same whatever the location of δm, and $\int \boldsymbol{\rho}dm = 0$ by definition of the centre of mass.

$$\begin{aligned}\int \mathbf{k}\wedge\mathbf{r}dm &= \int \mathbf{k}\wedge\mathbf{s}dm + \int \mathbf{k}\wedge\boldsymbol{\rho}dm \\ &= \mathbf{k}\wedge\mathbf{s}\int dm + \mathbf{k}\wedge\int\boldsymbol{\rho}dm \\ &= \mathbf{k}\wedge\mathbf{s}m\end{aligned}$$

So the last two terms in the force equation are

$$-m\mathbf{s}\dot{\theta}^2 + m\mathbf{k} \wedge \mathbf{s}\ddot{\theta} \tag{7.11}$$

They are shown as forces on the body in Fig. 7. 5. The forces on the *bearings* carrying the axis of rotation will be in the opposite directions.

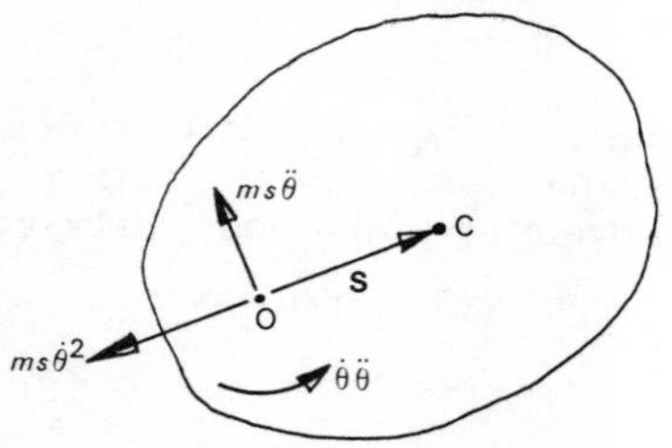

FIG. 7. 5

Summarising, for rotation about a fixed axis,

(a) $M_z = I_z\ddot{\theta}$, the moment driving the rotation.

(b) $H_z = I_z\dot{\theta}$, the moment of momentum.

(c) $M_z = \dot{H}_z = \dfrac{dT}{d\theta}$

(d) The time integral of the moment of the external forces is equal to the change of moment of momentum.

(e) The angle integral of the moment of the external forces is equal to the change of kinetic energy.

(f) The force at the axis equals the sum of all the external forces, $+ms\dot{\theta}^2$ along the line joining the axis to the centre of mass, $+ms\ddot{\theta}$ along a line at right angles, where s is the distance from the axis to the centre of mass.

Examples of the use of these will follow soon.

Moments of inertia

Integrals of the type $\int r^2 dm$ evaluated through the whole of a body are called moments of inertia and are often denoted by the symbol I, with a suffix to describe the axis about which I is measured. With reference to Fig. 7. 3, the appropriate axis is the z axis and so

$$I_z = \int r^2 dm$$

The value depends on the shape and density of the body, and on the location of the z axis, but it can easily be shown that:

The moment of inertia of a body about any axis equals the moment of inertia about the parallel axis through the centre of mass, plus the product of the mass of the body and the square of the distance between the two parallel axes.

With reference to Fig. 7.4,

$$\mathbf{r} = \mathbf{s} + \boldsymbol{\rho}$$

$$I_z = \int r^2 \mathrm{d}m$$

$$r^2 = \mathbf{r}\,.\,\mathbf{r} = (\mathbf{s} + \boldsymbol{\rho})\,.\,(\mathbf{s} + \boldsymbol{\rho})$$

$$= s^2 + 2\mathbf{s}\,.\,\boldsymbol{\rho} + \rho^2$$

So $I_z = s^2 \int \mathrm{d}m + 2\mathbf{s}\,.\int \boldsymbol{\rho}\mathrm{d}m + \int \rho^2 \mathrm{d}m$

(s is a constant)

The second term on the right-hand side is zero, because C is the centre of mass; the third term is the moment of inertia about an axis through the centre of mass. Therefore

$$I_z = I_C + ms^2$$

This is called the *parallel axes* theorem for moments of inertia. Note that it applies to parallel axes, *one of which passes through the centre of mass*, not to any two parallel axes.

The moment of inertia has dimensions mass x (length)^2, and is sometimes written

$$I = mk^2$$

where k is called the 'radius of gyration' of the body.

The moments of inertia of a number of simple bodies are given in an Appendix. These can be added together, using the parallel axes theorem where appropriate, to give values for simple composite bodies, such as a rotor with a number of wheels. But for complicated shapes one is usually driven to a numerical integration if an accurate value is required; or of course a measurement if the part has been made.

Examples of fixed axis rotation

Single bodies

Consider first the very simple case of a body rotating about an axis through its centre of mass, and accelerated by a weight hanging from a pulley on the same axis as the body. It is shown in Fig. 7.6 (a).

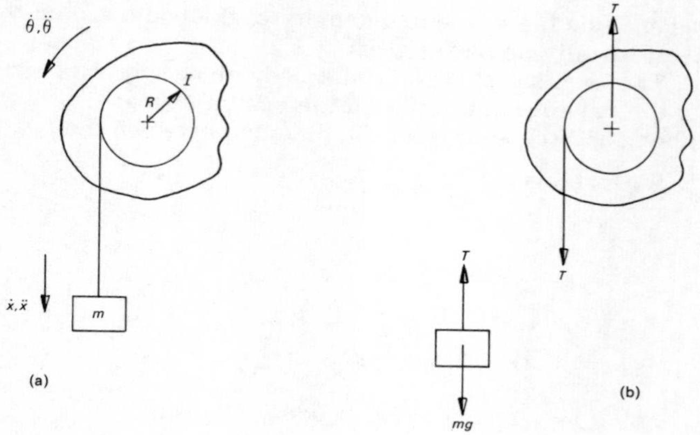

FIG. 7. 6

Suppose we want to know the velocity of the weight after it has descended a distance h from rest. The straightforward application of Newton's Second Law is as follows, with reference to the free-body diagrams in Fig. 7. 6 (b):

for the weight, $mg - T = m\ddot{x}$

$$T = mg - m\ddot{x}$$

for the rotor, $TR = I\ddot{\theta}$

$$I\ddot{\theta} = mgR - m\ddot{x}R$$

but $\ddot{x} = R\ddot{\theta}$

$$\ddot{\theta} = \frac{mgR}{I + mR^2}$$

and $\ddot{x} = \dfrac{g}{1 + I/mR^2}$

$$\ddot{x} = \dot{x}\,\frac{d\dot{x}}{dx}$$

integrating, $\tfrac{1}{2}\dot{x}^2 = \dfrac{g}{1 + I/mR^2}\,x + \text{constant}$

put $\dot{x} = 0$ at $x = 0$, then the velocity after fall h is given by

$$\dot{x}^2 = \frac{2gh}{1 + I/mR^2}$$

A much shorter solution is achieved using energy. Since the system as defined is conservative, we can use the conservation of (potential energy + kinetic energy):

gain of k.e. = loss of p.e.

$$\tfrac{1}{2} I \dot{\theta}^2 + \tfrac{1}{2} m \dot{x}^2 = mgh$$

$$\dot{x} = R\dot{\theta}$$

$$I \frac{\dot{x}^2}{R^2} + m\dot{x}^2 = 2mgh$$

$$\dot{x}^2 = \frac{2gh}{1 + I/mR^2}$$

Consider next another simple case, which is illustrative of many, less trivial practical cases, but without any geometrical complications. It is a straight beam or plate, hinged at one end to rotate about a horizontal axis, as in Fig. 7.7

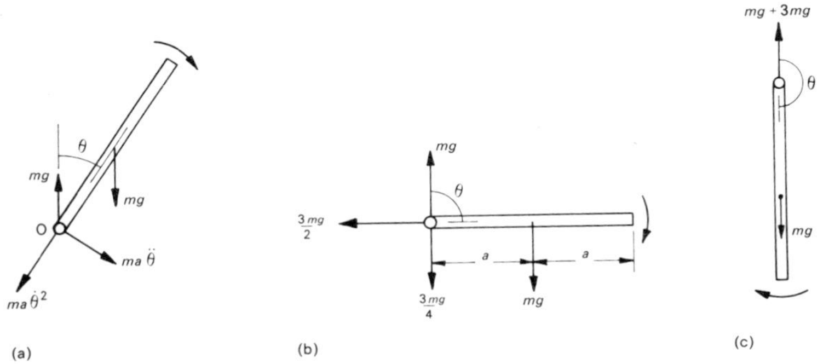

FIG. 7.7 The swinging beam

We set ourselves the task of finding the distribution of forces and bending moments in the beam as it swings. In a real situation the beam would be a part of a machine, and we would probably have to find its acceleration graphically or numerically by the methods we saw in Chapter 5.

Neglecting friction, the only forces acting on the beam are its own weight and the force at the hinge. We know the components of the force at the hinge from our general result (7.11); they are shown in Fig. 7.7 (a). The next step is to find the velocity and acceleration of the beam for any value of θ. To find the acceleration, we use

$$M_z = I_z \ddot{\theta}$$

In this case, $mga \sin\theta = I_0 \ddot{\theta}$

To find the velocity, we use energy, but we need to specify an initial value. Suppos the beam starts from rest at $\theta = 0$, and is given just a slight push to move it from the vertical (unstable) position. Then energy conservation gives

$$\tfrac{1}{2} I_0 \dot{\theta}^2 = mg(a - a\cos\theta)$$

Next we examine the way that forces and moments are transmitted along the beam (leading to a stress analysis in a real situation). To get the feel of the thing, we look at two particular positions, the horizontal ($\theta = \pi/2$) and the vertically downward ($\theta = \pi$). We shall then see that to do the general case is no more difficult.

In the horizontal position, $\sin\theta = 1$, $\cos\theta = 0$, so

$$\ddot{\theta} = \frac{mga}{I_0}, \dot{\theta}^2 = \frac{2mga}{I_0}$$

$$I_0 = \frac{ml^2}{3} = \frac{4ma^2}{3} \quad \text{(see Appendix)}$$

so $$\ddot{\theta} = \frac{3g}{4a}, \ \dot{\theta}^2 = \frac{3g}{2a}$$

The forces at the axis are therefore, with the arrows indicating direction,

$$mg\uparrow,\ ma\dot{\theta}^2 = \frac{3mg}{2}\leftarrow \text{ and } ma\ddot{\theta} = \frac{3mg}{4}\downarrow$$

They are shown on Fig. 7. 7 (b). Note that the net vertical force $\frac{mg}{4}\uparrow$ on the beam at the axis depends on the value of $\ddot{\theta}$ and is therefore *independent* of the value of θ at release; whereas the $\frac{3mg}{2}\leftarrow$ depends on the value of $\dot{\theta}$ and so on the release poi also.

Figure 7. 7 (b) shows the *external* forces on the beam. To find the internal forces we must isolate an element of the beam. Figure 7. 8 shows such a element and the forces acting on it. T is the tension in the beam, F the shear forc and M the bending moment. These can all be expected to vary along the length, so small changes are shown between the ends of the element.

The weight of the element is $\frac{m}{l}\,\delta r g$, m being the mass of the whole beam. Finally, in dashed lines, are shown the 'd'Alembert forces'. (Note that no d'Alembert coup is shown, because the element has a moment of inertia which is of 'the second order of small quantities'.) The equations of motion of the element are therefore,

$$\rightarrow T + \delta T + \frac{m}{l} r\dot{\theta}^2 \delta r - T = 0$$

$$\uparrow F + \frac{m}{l} r\ddot{\theta}\delta r - \frac{m}{l}\,\delta r g - F - \delta F = 0$$

$$\curvearrowleft M - M - \delta M + F\frac{\delta x}{2} + (F + \delta F)\frac{\delta x}{2} = 0$$

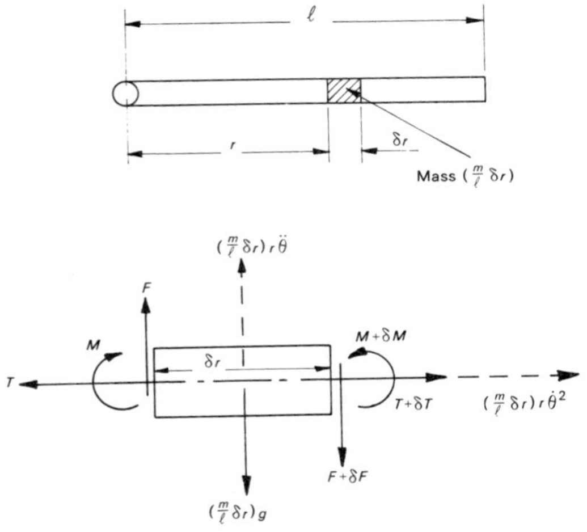

FIG. 7. 8

As $\delta r \rightarrow 0$, these give

$$\frac{\mathrm{d}T}{\mathrm{d}r} = -\frac{m}{l} r\,\dot{\theta}^2 = -\frac{3mg}{l^2} r \tag{7.12}$$

$$\frac{\mathrm{d}F}{\mathrm{d}r} = \frac{m}{l}(r\ddot{\theta} - g) = \frac{mg}{l}\left(\frac{3r}{2l} - 1\right) \tag{7.13}$$

$$\frac{\mathrm{d}M}{\mathrm{d}r} = F \tag{7.14}$$

Solving these in turn,

(7. 12) $$T = -\frac{3mg}{l^2}\frac{r^2}{2} + \text{constant}$$

Plainly $T = 0$ at $r = l$, so constant $= \frac{3mg}{2}$

$$T = \frac{3mg}{2}\left(1 - \frac{r^2}{l^2}\right)$$

Note that at $r = 0$, $T = \frac{3mg}{2}$, which is in accord with Fig. 7. 7 (b)

(7. 13) $$F = \frac{mg}{l}\left(\frac{3}{2l}\cdot\frac{r^2}{2} - r\right) + \text{constant}$$

$F = 0$ at $r = l$, so constant $= \frac{mg}{4}$

$$F = \frac{mg}{4}\left(1 - \frac{4r}{l} + \frac{3r^2}{l^2}\right)$$

At $r = 0$, $F = mg$, as required by Fig. 7.7 (b).

(7.14) $$M = \int F\mathrm{d}r$$

$$= \frac{mg}{4}\left(r - \frac{2r^2}{l} + \frac{r^3}{l^2}\right) + \text{constant}$$

Since one end of the beam is free, and the other is hinged, M must be zero at both ends.

$$M = 0 \text{ at } r = 0 \text{ gives constant} = 0$$

At $$r = l,\ M = \frac{mg}{4}(l - 2l + l) = 0$$

So $$M = \frac{mgl}{4}\left(\frac{r}{l} - 2\frac{r^2}{l^2} + \frac{r^3}{l^3}\right)$$

The variations of M, F and T along the length are shown in Fig. 7.9. In the vertical position, $\theta = \pi$,

$$\ddot{\theta} = 0,\ \dot{\theta}^2 = \frac{4mga}{I_0} = \frac{3g}{a} = \frac{6g}{l}$$

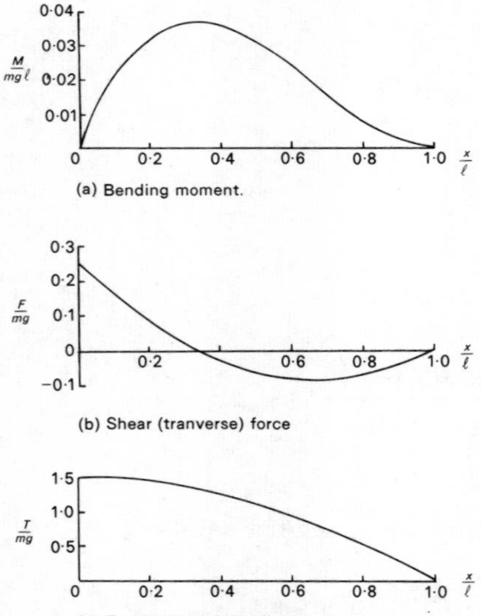

FIG. 7.9 Forces in swinging beam, $\theta = \pi/2$

The forces at the axis are therefore,

$$mg\uparrow,\ ma\dot{\theta}^2 = 3mg\uparrow \text{ and } ma\ddot{\theta} = 0\leftarrow$$

They are shown on Fig. 7. 7 (c).

Again we isolate an element δr, Fig. 7. 10. Assure yourself that all the forces are shown.

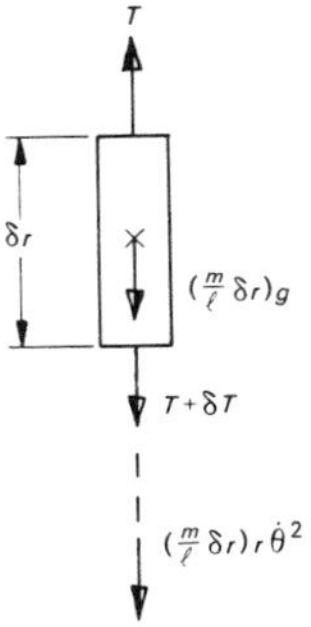

FIG. 7. 10

The equation of motion is

$$T + \delta T + \frac{m}{l}\delta r g + \frac{m}{l}\delta r \,.\, r\dot{\theta}^2 - T = 0$$

$$\frac{dT}{dr} = -\frac{m}{l}(g + r\dot{\theta}^2) = -\frac{mg}{l}\left(1 + \frac{6r}{l}\right)$$

$$T = -\frac{mg}{l}\left(r + \frac{3r^2}{l}\right) + \text{constant}$$

$$T = 0 \text{ at } r = l, \text{ constant} = 4mg$$

$$T = mg\left(4 - \frac{r}{l} - \frac{3r^2}{l^2}\right)$$

Before leaving this problem, write down the equations of motion for an element when the beam is at a general angle of inclination θ. Satisfy yourself that you could complete the analysis.

Connected bodies

Systems of connected bodies, all rotating about fixed axes, are of necessity simple, in principle anyway. The fixed axis proviso rules out the simple four-bar chain, the engine mechanism and epicyclic gear trains. It leaves bodies connected by ordinary gear trains, belt-drives and various friction devices such a clutches.

Consider first two rotors, connected by gears or a chain drive, or an positive transmission system, as in Fig. 7. 11. To find the acceleration due to a

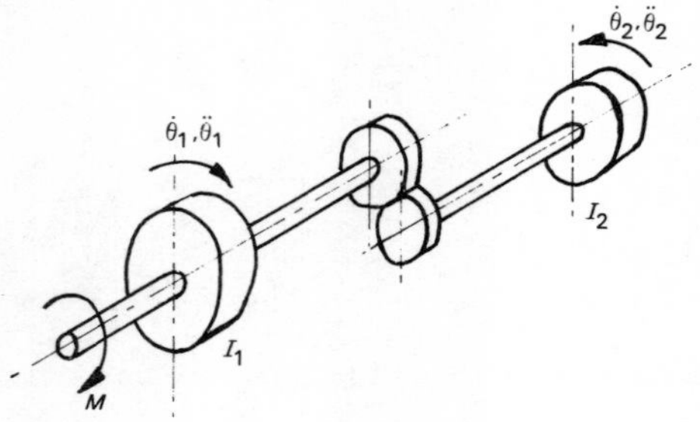

FIG. 7. 11 Geared rotors

given driving torque M, isolate the two rotors and examine the forces at the gear wheels. Figure 7. 12 shows the forces on the two rotors, other than those at the axes. P and Q are the tangential and radial components of the force between the gear wheels. Applying moment/angular acceleration to the two rotors gives

$$M - PR_1 = I_1\ddot{\theta}_1$$

$$PR_2 = I_2\ddot{\theta}_2$$

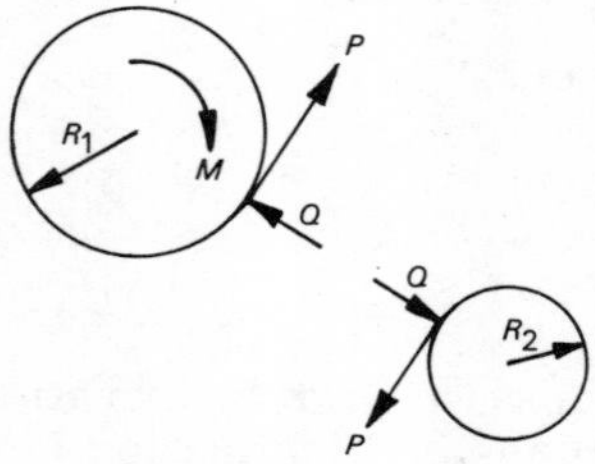

FIG. 7. 12 Forces on rotors

If the gear ratio, driven speed: driver speed, is G, then

$$\dot{\theta}_2 = G\dot{\theta}_1 \text{ and } \ddot{\theta}_2 = G\ddot{\theta}_1$$

The gears have the same pitch-circle speed, so

$$R_1\dot{\theta}_1 = R_2\dot{\theta}_2, \quad \frac{R_1}{R_2} = G$$

Solving the equations of motion

$$M = I_1\ddot{\theta}_1 + PR_1$$

$$PR_1 = PR_2 \cdot \frac{R_1}{R_2} = I_2\ddot{\theta}_2\frac{R_1}{R_2} = I_2G^2\ddot{\theta}_1$$

So
$$M = (I_1 + G^2I_2)\ddot{\theta}_1 \qquad (7.15)$$

The G^2 shows the extent of the burden of an additional rotor, and gives a clue to the success of small toys with flywheels, such as 'push-and-go' motor cars.

Show that the kinetic energy of the rotors can be written

$$T = \tfrac{1}{2}(I_1 + G^2I_2)\dot{\theta}_1^{\,2}$$

and extend the expressions to cover any number of rotors geared together, but all with fixed axes.

General plane motion

It will be shown later in this chapter, as a special case of general motion of a body, that for motion in a plane with translation and rotation, the following simple results apply. For motion in the xy plane ($z = 0$), the motion of the centre of mass of the body is determined by

$$\mathbf{F} = m\mathbf{a} \qquad (7.16)$$

where **F** is the resultant of all the forces acting on the body and **a** is the acceleration of the centre of mass. In cartesian components

$$F_x = m\ddot{x}$$
$$F_y = m\ddot{y} \qquad (7.17)$$

In addition there is a rotational motion about the centre of mass governed by

$$M = I_{\text{C}}\ddot{\theta} \qquad (7.18)$$

where M is the total moment about the centre of mass of all the forces acting on the body, and I_{C} is the moment of inertia about the axis through the centre of mass and perpendicular to the plane of motion.

The importance of referring all these quantities to the centre of mass cannot be over-emphasised.

The relations involving *momentum and energy* are as follows.

The resultant of the forces applied to the body equals the rate of change of momentum of the body regarded as a particle at the centre of mass.

The total moment of all the forces about a point 0 in the plane of motion is equal to the rate of change of *moment of momentum* of the body about 0. This requires elaboration.

First, (the *moment of momentum* about 0) is (the moment of moment about C) + (the moment about 0 of the momentum of the body regarded as a partic at the centre of mass). With reference to Fig. 7.13, the moment of momentum

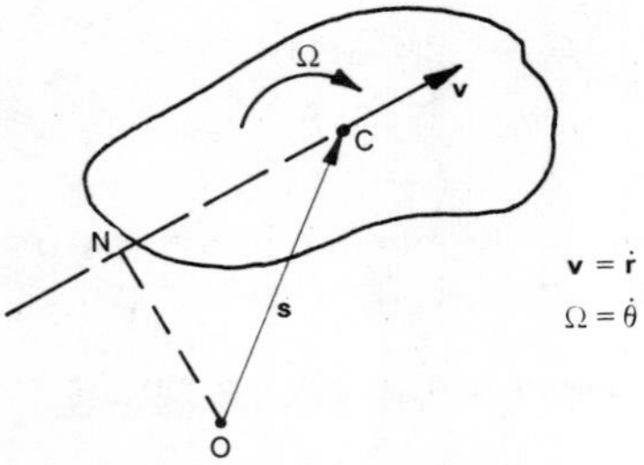

FIG. 7.13 Moment of momentum

about 0,

$$\mathbf{H_O} = \mathbf{H_C} + \mathbf{s} \wedge m\mathbf{v}$$

or $$H_O = I_C\,\Omega + (\mathrm{ON} \times mv) \qquad (7.1$$

since all the vectors are perpendicular to the *xy* plane.

Second, the location of 0 : 0 can be any fixed point in the plane of motion, or any point moving with a velocity parallel to that of the centre of mass, including of course the centre of mass itself.

If 0 is chosen on the line of the resultant force, then there will be no change in the moment of momentum about 0. This choice is particularly usefu in impacts, where the impact force is unknown; typical applications are the mathe maticians' delights, the rough, inelastic hoop striking a kerb; and the cricket bat problem, where should the ball strike the bat in order not to sting the batsman's hands?

In the special case of the body *rotating about* 0, then $v = s\Omega$ and

$$\begin{aligned} H_O &= I_C\Omega + s\,(ms\Omega) \\ &= (I_C + ms^2)\,\Omega \\ &= I_O\Omega \end{aligned}$$

which we have seen before.

The kinetic energy of the body is (the kinetic energy of rotation abo the centre of mass) + (the kinetic energy of the body regarded as a particle at th centre of mass)

$$T = {}^1/_2\, I_C\Omega^2 + {}^1/_2\, mv^2 \qquad (7.2$$

Inertia effects in machines

The effects of the mass and moment of inertia of a part in a machine can usually be found most easily by direct application of equations (7.16)-(7.18), using (7.16) in a graphical solution and probably (7.17) in an analytical solution. It is wise to consider the links in a machine separately; in other words to isolate them individually. It is also useful to show (7.16) and (7.18) as 'd'Alembert force and couple' on a free-body diagram, referred to the centre of mass; then it is in order to take moments about any point in the plane of motion.

Whereas in propulsion and traction applications one is usually seeking the motion resulting from known forces, it is usual in machines for the motion to be prescribed, and the designer is then faced with the problem of finding the forces in the various parts, so that he can decide how thick and of what to make them. Of course the forces depend to some extent on these things and so there may be a good deal of trial and error in the design stage. It should be added that in order to prescribe the motion it is usual to include a fairly large inertia, such as a flywheel, to maintain practically constant angular velocities of rotating parts. By such means speed fluctuations are kept within reasonable limits even though the drive may be irregular, as in a reciprocating engine; or where the power demand is widely varying, as in all sorts of pressing machines; or where there are heavy oscillating parts, again as in a reciprocating engine.

Figure 7.14 shows a link AB in a machine whose motion is known and is such that, in the position considered, the acceleration of the centre of mass is **a** and the angular acceleration of the link is $\ddot{\theta}$ as in (a). The d'Alembert force and couple are shown in (b), and have been combined in (c) by shifting the force parallel to itself by a distance d, given by

$$mad = I_C \ddot{\theta}$$

reducing the force and couple to a single force. In a graphical solution it is useful to calculate d and draw a circle of radius d with centre at the centre of mass; then it is easy to construct the shifted $m\mathbf{a}$ as a tangent to the circle.

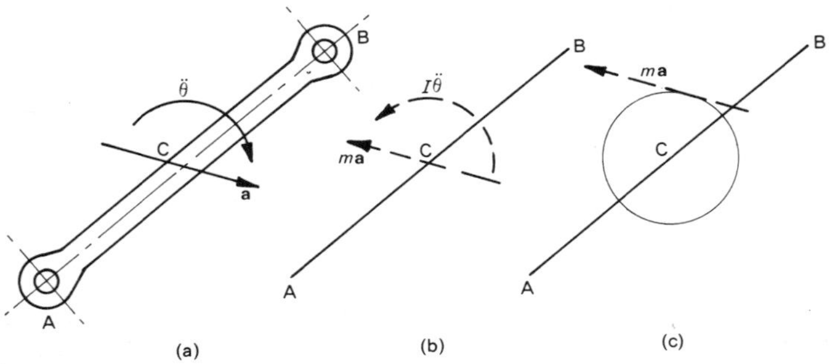

FIG. 7.14 Inertia effects in a machine link

Since *real* forces can only be applied to the link at A and B (apart from its own weight), the link has now been simplified to a 'three-force member'. We can therefore work out what real external forces are required to give the link its accelerations. Note that this will give us only the *external* forces; to find the whole stress pattern inside the link we should have to give it the full 'swinging beam' treatment, but first we would be well advised to find out the order of magnitude of the d'Alembert forces compared with the external forces driving the mechanism, to make sure whether the elaborate calculation was worth-while or not.

As an example, let us consider an engine mechanism, and isolate the connecting rod to find what forces are required at its ends to accelerate it, for a given engine position and speed. These forces could then be superimposed on any other forces being transmitted, but which we are not considering.

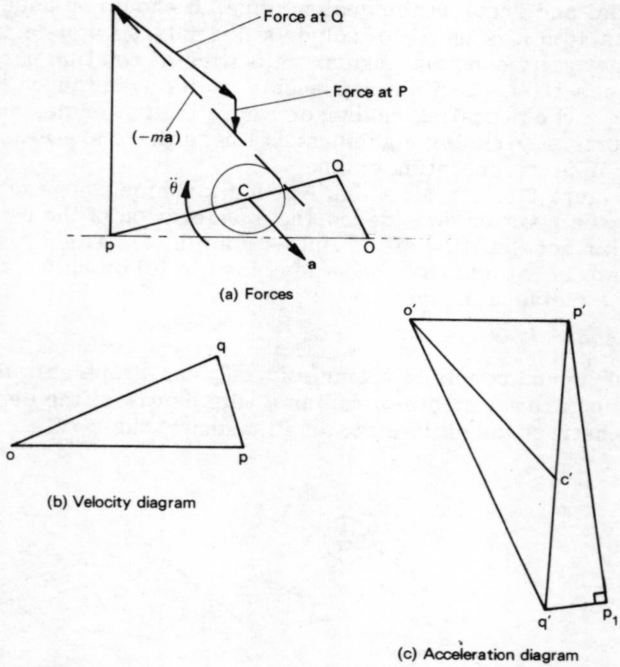

FIG. 7.15

Figure 7.15 (a) shows the position of the connecting rod PQ to be considered. (b) and (c) are the velocity and acceleration diagrams. Recalling briefly the details from Chapter 5, oq is the vector representing the known velocity of Q to some suitable scale, drawn perpendicular to OQ; qp is the velocity of P relative to Q and is drawn perpendicular to PQ, to give the location of p by intersection with the known direction of the velocity of P. The acceleration diagram is drawn by first constructing o′q′, the acceleration of Q (centripetal only), then q′p,, the centripetal component of the acceleration of P relative to Q, followed by p,p′ at right angles to q′p, giving p′ by intersection with o′p′. The acceleration of the centre of mass is found by joining p′q′ and dividing its length at c′ in the same ratio as PQ is divided by C. Then o′c′ gives the value of **a**. The angular acceleration of the connecting rod, $\ddot{\theta}$, is given by

$$\ddot{\theta} = \frac{p'p,}{PQ}$$

p′p, being the tangential component of the acceleration of Q relative to P. The shift of ($-m\mathbf{a}$) required to combine it with the d'Alembert couple ($-I_C\ddot{\theta}$) is

$$d = \frac{I_C\ddot{\theta}}{ma}$$

and a circle of this radius is drawn round the centre of mass in Fig. 7.15 (a). Finally a triangle of forces is drawn, based on the d'Alembert force ($-m\mathbf{a}$) shown in its true line of action. The real forces at P and Q must intersect with the d'Alembert force, and the point of intersection has been found by assuming that the friction between the piston and cylinder is negligible and so the force at P (to accelerate the connecting rod) is perpendicular to the stroke.

The connecting rod will transmit other forces, due to the gas-pressure on the piston, and the piston's own d'Alembert force, but they can be added to the forces we have just found.

It is now clear that to do the whole job accurately for a number of positions of the mechanism would be a very time-consuming business. We can however analyse the engine mechanism approximately in a very simple way. The connecting rod is the only part which is not simply rotating or reciprocating. If we could replace it dynamically by two masses, one at P and one at Q, we would then have only reciprocating and rotating parts, connected by a massless link PQ. Can it be done?

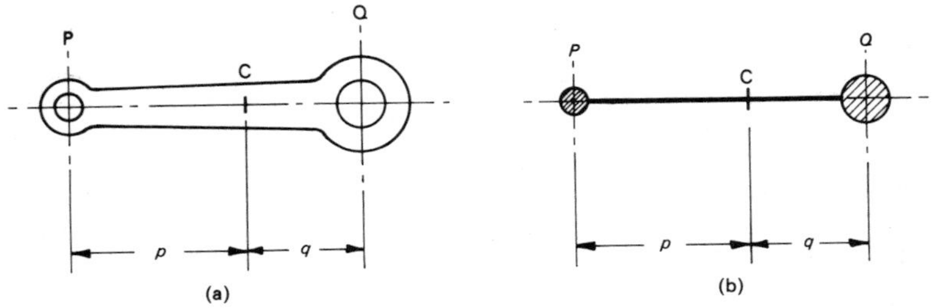

FIG. 7. 16 Connecting rod

A two-mass equivalent of the connecting rod must have the same total mass, the same centre of mass, and the same moment of inertia. If the rod of Fig. 7.16 (a) is to be replaced by the two masses P and Q of (b), there are three conditions to be satisfied and only two variables, the masses P and Q. This simple model will not, therefore, be the dynamical equivalent of the real thing. It is usual to accept a statical equivalent (same total mass and position of centre of mass), and sacrifice the moment of inertia.

Same mass $\quad P + Q = m$

Same centre of mass $\quad pP = qQ$

Therefore $\quad P = \dfrac{q}{p+q}m, \; Q = \dfrac{p}{p+q}m$

The moment of inertia of the two masses will be

$$Pp^2 + Qq^2$$

treating the masses as particles.

This equals $\quad m\left[\dfrac{p^2 q}{p+q} + \dfrac{q^2 p}{p+q}\right] = mpq$

Full dynamical equivalence would only be achieved if

$$mpq = I_C = mk^2{}_C \quad \text{the value for the connecting rod.}$$

The mass of the connecting rod is distributed, so $k^2{}_C$ will be somewhat less than pq. In other words the simple approximation will slightly *overestimate* the effect of $I_C\ddot{\theta}$.

Now in the engine mechanism the mass Q will be added to that of the crank—a purely rotating mass—and could quite easily be balanced dynamically by the addition of further rotating masses. The mass P is added to that of the piston, making a total reciprocating mass R, say.

We saw in Chapter 5, equation (5.6), that a good approximation to the piston acceleration is,

$$\ddot{x} = r\omega^2\left(\cos\theta + \frac{\cos 2\theta}{n}\right)$$

where r is the crank radius and n is ratio of connecting rod length to crank radius; θ is the crank angle measured from zero at 'top-dead-centre'. $\ddot{x}$ is clearly a maximum at $\theta = 0$ when both the cosine terms are +1, so the maximum force required to accelerate the reciprocating mass is

$$Rr\omega^2\left(1 + \frac{1}{n}\right)$$

To get an idea of the order of this, consider a small motor-car engine, with four cylinders and the following dimensions:

connecting rod	140 mm long $\left.\begin{array}{l} P = 0.17 \text{ kg} \\ Q = 0.33 \text{ kg} \end{array}\right\}$ 0.5 kg total
piston	60 mm diameter 0.25 kg mass
crank radius	35 mm
maximum piston acceleration:	
at 1 000 rev/min	($\omega \approx 105$ rad/s) maximum
acceleration =	480 m/s^2
at 5 000 rev/min	maximum
acceleration =	1.2×10^4 m/s^2
R =	$0.25 + 0.17 = 0.42$ kg

So the force required to accelerate one piston is approximately

200 N at 1 000 rev/min
5 000 N at 5 000 rev/min

At high speeds, therefore, this force is comparable with the maximum gas force, which would be of the order

$$\frac{\pi}{4} \times 0.06^2 \times 5 \times 10^6 \text{ N} = 14\,000 \text{ N}$$

for a maximum gas pressure of the order of 5×10^6 N/m^2 (50 atmospheres). And of course the *mean* gas force over the whole working cycle would be very much less than the 50 atmospheres used in the example.

So the forces required to accelerate the reciprocating parts emerge as major forces in high-speed piston engines. But the fact that these forces are comparable with the gas forces is practically irrelevant as far as power output is concerned, because these 'd'Alembert forces' are essentially conservative; drawing on the energy in the flywheel over part of the cycle, and returning energy to the flywheel over the remainder of the cycle; and in multi-cylinder engines, in part cancelling each other out.

Although the d'Alembert forces do not present a power-loss problem, they do cause vibration and balancing problems, due to their failure to cancel out completely, even in multi-cylinder configurations. Engine balancing has attracted much attention and ingenuity, and is in itself an interesting and fairly long story—too long to be told here, but summarised in [1].

General Dynamics

The next few pages contain proofs of some earlier assertions, a few additional results and one or two applications. The first few results to be derived apply to any system of particles, including of course a rigid body, but including also any non-rigid system, viz. deformable bodies or fluids. The brief outline analysis will concentrate on considerations of momentum and the equations of motion of a rigid body. A much fuller discussion is given in [2].

System of particles

The equation of motion of a particle δm at $\mathbf{r}$ in a system is

$$\mathbf{F} + \mathbf{f} = \delta m\ddot{\mathbf{r}}$$

where $\mathbf{F}$ is the resultant external force and $\mathbf{f}$ the resultant internal force on the particle, as understood in Chapter 4 (Fig. 4. 2) and earlier in this chapter. For the whole system, all the $\mathbf{f}$'s will cancel, and so

$$\Sigma\,\mathbf{F} = \Sigma\,\delta m\ddot{\mathbf{r}}$$

Now the position of the centre of mass of the system is defined by the vector s, such that

$$m\mathbf{c} = \Sigma\,\delta m\mathbf{r}$$

where $m = \Sigma\,\delta m$, the mass of the whole system. Differentiating twice with respect to time clearly gives

$$\Sigma\,\mathbf{F} = m\ddot{\mathbf{c}} \tag{7.21}$$

So the centre of mass of a system of particles moves as if all the mass were concentrated there, and the resultant of all the external forces acted there.

Moment of momentum

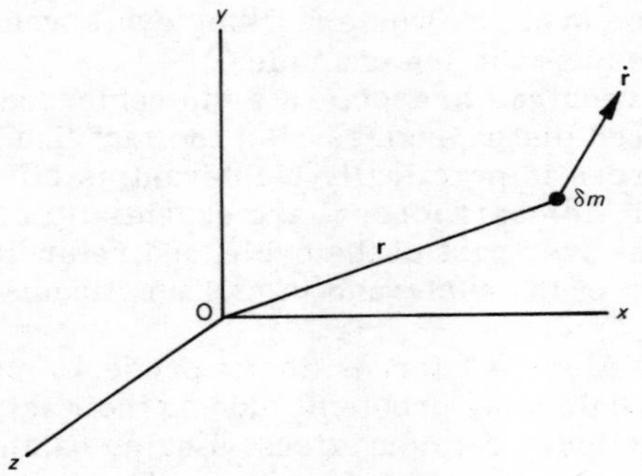

FIG. 7. 17

The momentum of the particle δm in Fig. 7.17 is $\delta m\dot{\mathbf{r}}$. The moment of momentum of the particle about 0 is $\mathbf{r} \wedge \delta m\dot{\mathbf{r}}$. The total moment of momentum of the system about 0 is

$$\Sigma\mathbf{r} \wedge \delta m\dot{\mathbf{r}} = \mathbf{H}_O$$

Differentiating with respect to time gives

$$\Sigma\dot{\mathbf{r}} \wedge \delta m\dot{\mathbf{r}} + \Sigma\mathbf{r} \wedge \delta m\ddot{\mathbf{r}} = \dot{\mathbf{H}}_O$$

Now $\dot{\mathbf{r}} \wedge \dot{\mathbf{r}} = 0$, and $\delta m\ddot{\mathbf{r}} = \mathbf{F} + \mathbf{f}$

so $\Sigma\mathbf{r} \wedge (\mathbf{F} + \mathbf{f}) = \mathbf{H}_O$

The left-hand side represents the moments of all the forces about O; since the internal forces cancel, so also do their moments about O. So the left-hand side equals the sum of the moments of all the external forces about O:

$$\mathbf{M}_O = \dot{\mathbf{H}}_O \qquad (7.22)$$

$\mathbf{H}_O$ is the moment of momentum about a fixed point, but the same result holds for certain *moving* points.

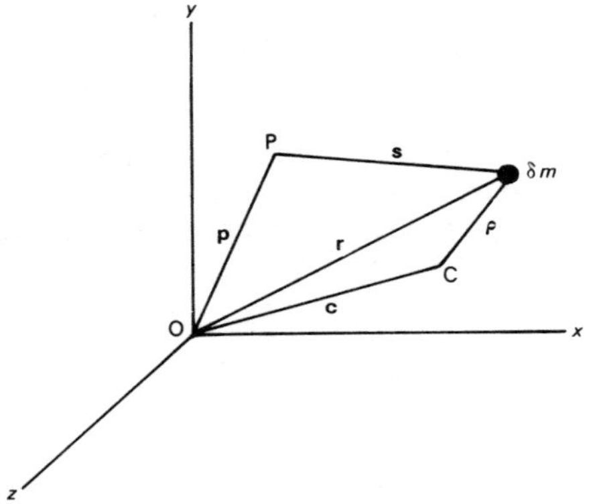

FIG. 7.18

Referring to Fig. 7.18, the moment of momentum of δm about the moving point P is $\mathbf{s} \wedge \delta m\dot{\mathbf{r}}$, so for the system of particles

$$\Sigma\mathbf{s} \wedge \delta m\dot{\mathbf{r}} = \mathbf{H}_P$$

Differentiating with respect to time

$$\Sigma\dot{\mathbf{s}} \wedge \delta m\dot{\mathbf{r}} + \Sigma\mathbf{s} \wedge \delta m\ddot{\mathbf{r}} = \dot{\mathbf{H}}_P$$

$$\mathbf{r} = \mathbf{p} + \mathbf{s} = \mathbf{c} + \boldsymbol{\rho}$$

So $\quad \dot{\mathbf{s}} = \dot{\mathbf{r}} - \dot{\mathbf{p}}$ and $\dot{\mathbf{r}} = \dot{\mathbf{c}} + \dot{\boldsymbol{\rho}}$

Therefore $\quad \Sigma\dot{\mathbf{s}} \wedge \delta m\dot{\mathbf{r}} = -\dot{\mathbf{p}} \wedge \Sigma\delta m\dot{\mathbf{r}}$

$$= -\dot{\mathbf{p}} \wedge \dot{\mathbf{c}}\,\Sigma\,\delta m - \dot{\mathbf{p}} \wedge \Sigma\dot{\boldsymbol{\rho}}\,\delta m$$

$$= -m\dot{\mathbf{p}} \wedge \dot{\mathbf{c}}$$

$$\left(\Sigma\,\dot{\boldsymbol{\rho}}\delta m = \frac{\mathrm{d}}{dt}\,\Sigma\boldsymbol{\rho}\,\delta m = 0 \text{ by definition of centre of mass}\right)$$

Also $\quad \Sigma\,\mathbf{s} \wedge\ \delta m\ddot{\mathbf{r}} = \ \Sigma\,\mathbf{s} \wedge (\mathbf{F} + \mathbf{f}) = \mathbf{M}_{\mathrm{P}}$

the moment about P of all the external forces.

So we have

$$\mathbf{M}_{\mathrm{P}} - m\dot{\mathbf{p}} \wedge \dot{\mathbf{c}} = \dot{\mathbf{H}}_{\mathrm{P}}$$

$\dot{\mathbf{p}}$ is the velocity vector of P and $\dot{\mathbf{c}}$ is the velocity vector of the centre of mass C. *If they are parallel*, $\dot{\mathbf{p}} \wedge \dot{\mathbf{c}} = 0$, or of course if $\dot{\mathbf{p}} = 0$,

$$\mathbf{M}_{\mathrm{P}} = \dot{\mathbf{H}}_{\mathrm{P}} \tag{7.23}$$

This expression holds then for any point P which

(1) is fixed, (2) is the centre of mass, or (3) is moving with a velocity parallel to that of the centre of mass.

Within these restrictions on P, if the component in any direction of the resultant moment of the external forces about P is zero, then the component of the moment of momentum about P in that direction is conserved.

Returning for a moment to

$$\mathbf{H}_{\mathrm{P}} = \Sigma\,\mathbf{s} \wedge \delta m\dot{\mathbf{r}}$$

$$\mathbf{s} \ = \mathbf{r} - \mathbf{p} \ \text{ and } \ \dot{\mathbf{r}} = \dot{\mathbf{c}} + \dot{\boldsymbol{\rho}}$$

$$\mathbf{H}_{\mathrm{P}} = \Sigma\mathbf{r} \wedge \delta m\dot{\mathbf{r}} - \mathbf{p} \wedge \Sigma\,\delta m\dot{\mathbf{r}}$$

$$= \mathbf{H}_{\mathrm{O}} - \mathbf{p} \wedge m\dot{\mathbf{c}} - \mathbf{p} \wedge \Sigma\,\dot{\boldsymbol{\rho}}\delta m$$

The last term is zero by definition of the centre of mass. Therefore

$$\mathbf{H}_{\mathrm{O}} = \mathbf{H}_{\mathrm{P}} + \mathbf{p} \wedge m\dot{\mathbf{c}}$$

In particular, if P is at the centre of mass, $\mathbf{p} = \mathbf{c}$, and

$$\mathbf{H}_{\mathrm{O}} = \mathbf{H}_{\mathrm{C}} + \mathbf{c} \wedge m\dot{\mathbf{c}} \tag{7.24}$$

In words, the moment of momentum of a system about a fixed point O is equal to the moment of momentum of the system about the centre of mass, plus the moment of momentum about O of a particle whose mass is equal to the mass of the system and is located at the centre of mass.

Kinetic energy

Finally, for a system of particles, the kinetic energy can be referred to the centre of mass.

$$T = \Sigma\, {}^1\!/_2 \delta m v^2$$

$$\mathbf{r} = \mathbf{c} + \boldsymbol{\rho}$$

$$v^2 = \dot{\mathbf{r}} \,.\, \dot{\mathbf{r}} = (\dot{\mathbf{c}} + \dot{\boldsymbol{\rho}}) \,.\, (\dot{\mathbf{c}} + \dot{\boldsymbol{\rho}})$$

$$= \dot{c}^2 + 2\dot{\mathbf{c}} \,.\, \dot{\boldsymbol{\rho}} + \dot{\rho}^2$$

Therefore $$T = \Sigma\, {}^1\!/_2 \delta m \dot{c}^2 + \dot{\mathbf{c}} \,.\, \Sigma \dot{\boldsymbol{\rho}} \delta m + \Sigma\, {}^1\!/_2 \delta m \dot{\rho}^2$$

$$T = {}^1\!/_2 m \dot{c}^2 + \Sigma\, {}^1\!/_2 \delta m \dot{\rho}^2 \tag{7.25}$$

In words, the kinetic energy of a system of particles is equal to the kinetic energy of a particle of mass m with velocity equal to that of the centre of mass, plus the kinetic energy of the system corresponding to its motion relative to the centre of mass.

Summarising, the results (7.21)-(7.25) apply to any system of particles

(7.21) $\Sigma\, \mathbf{F} = m\ddot{\mathbf{c}}$

(7.22) $\mathbf{M}_O = \dot{\mathbf{H}}_O$

which is a special case of

(7.23) $\mathbf{M}_P = \dot{\mathbf{H}}_P$

where P is a point which is either (1) fixed, (2) the centre of mass or (3) moving with a velocity parallel to that of the centre of mass.

(7.24) $\mathbf{H}_O = \mathbf{H}_C + \mathbf{c} \wedge m\dot{\mathbf{c}}$
$= \mathbf{H}_C + \mathbf{c} \wedge m\boldsymbol{v}_C$

(7.25) $T = {}^1\!/_2 m \dot{c}^2 + \Sigma\, {}^1\!/_2 \delta m \dot{\rho}^2$

Rigid bodies

We start by defining the motion of a rigid body. We do this by fixing to the body a set of xyz axes, and allowing those axes to move within a fixed XYZ frame, as we did in the later part of Chapter 5. We define the motion of the xyz axes as before, by describing the origin as $\boldsymbol{R}$ in the XYZ axes, and the rotational motion as $\boldsymbol{\omega}$.

Any point in the body is fixed in the xyz frame, because the body is rigid, and its velocity is given by equation (5.8)

$$\dot{\mathbf{r}} = \dot{\mathbf{R}} + \boldsymbol{\omega} \wedge \boldsymbol{\rho} \tag{7.26}$$

($\dot{\rho}_r = 0$ because the body is fixed in the xyz axes)

The motion of the centre of mass of the body is determined by (7.21) Our attention is concentrated on the equations governing the rotation.

Moment of momentum

Equation (7.23) relates the moment of the external forces to the rate of change of moment of momentum about a variety of points.

$$\mathbf{M} = \dot{\mathbf{H}}$$

A useful start is to find an expression for **H** about the centre of mass, $\mathbf{H}_C$. To do this we take the origin of *xyz* coordinates at the centre of mass, as in

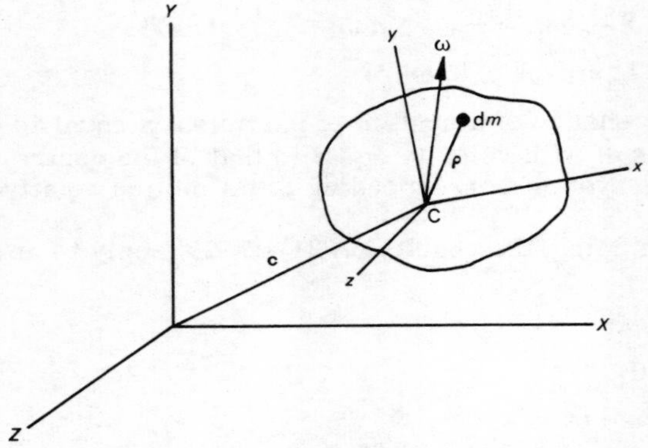

FIG. 7.19

Fig. 7.19. The *xyz* axes are fixed to the body with origin at the centre of mass, and rotate with angular velocity $\boldsymbol{\omega}$. d*m* is an element of mass, at $\boldsymbol{\rho}$ from the centre of mass, with velocity $\boldsymbol{v}$. The moment of momentum of d*m* about c is

$$\boldsymbol{\rho} \wedge \mathrm{d}m\boldsymbol{v}$$

The total for the body is

$$\mathbf{H}_C = \int (\rho \wedge \boldsymbol{v})\, \mathrm{d}m$$

From (7.26),

$$\begin{aligned}\mathbf{v} &= \dot{\mathbf{R}} + \boldsymbol{\omega} \wedge \boldsymbol{\rho}\\ &= \dot{\mathbf{c}} + \boldsymbol{\omega} \wedge \boldsymbol{\rho} \quad \text{in this case}\end{aligned}$$

$$\begin{aligned}\mathbf{H}_C &= \int \boldsymbol{\rho} \wedge (\dot{\mathbf{c}} + \boldsymbol{\omega} \wedge \boldsymbol{\rho})\, \mathrm{d}m\\ &= \int \boldsymbol{\rho} \wedge \dot{\mathbf{c}}\, \mathrm{d}m + \int \boldsymbol{\rho} \wedge (\boldsymbol{\omega} \wedge \rho)\, \mathrm{d}m\end{aligned}$$

$\dot{\mathbf{c}}$ is the same for all points in the body, so

$$\int \boldsymbol{\rho} \wedge \dot{\mathbf{c}}\, \mathrm{d}m = \left(\int \boldsymbol{\rho}\, \mathrm{d}m\right) \wedge \dot{\mathbf{c}} = 0$$

So $$\mathbf{H}_C = \int \boldsymbol{\rho} \wedge (\boldsymbol{\omega} \wedge \boldsymbol{\rho})\, dm \tag{7.27}$$

Equation (7.24) gives the value of **H** about a fixed point in terms of $\mathbf{H}_C$, for any system of particles, including a rigid body. A similar result can be derived for a rigid body which applies to any point which is fixed *relative to the body*, i.e. fixed in the *xyz* axes.

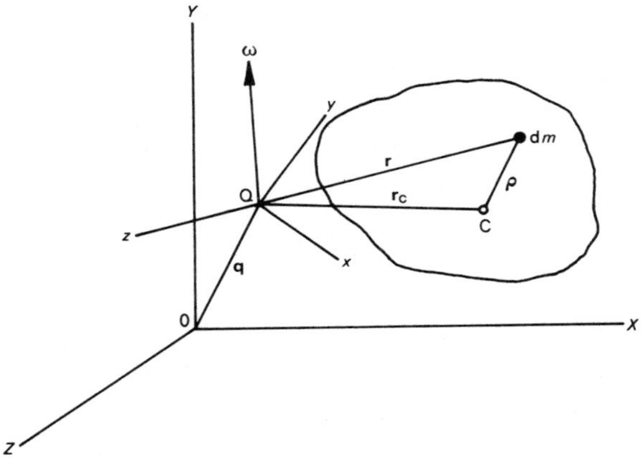

FIG. 7.20

In Fig. 7.20 the point Q is fixed in the *xyz* axes; in fact it is taken as the origin of the *xyz* axes. It is therefore 'attached' to the body, although not necessarily inside it.

The moment about Q of the momentum of the element dm is

$$\mathbf{r} \wedge dm\, v$$

so the total for the body is

$$\mathbf{H}_Q = \int (\mathbf{r} \wedge v)\, dm$$

$$v = \dot{\mathbf{q}} + \boldsymbol{\omega} \wedge \mathbf{r}, \quad \mathbf{r} = \mathbf{r}_C + \boldsymbol{\rho}$$

$$\begin{aligned}\mathbf{H}_Q &= \int (\mathbf{r} \wedge \dot{\mathbf{q}})\, dm + \int \mathbf{r} \wedge (\boldsymbol{\omega} \wedge \mathbf{r})\, dm \\ &= \int \mathbf{r}_C \wedge \dot{\mathbf{q}})\, dm + \int (\boldsymbol{\rho} \wedge \dot{\mathbf{q}})\, dm \\ &\quad + \int \mathbf{r}_C \wedge (\boldsymbol{\omega} \wedge \mathbf{r}_C)\, dm + \int \mathbf{r}_C \wedge (\boldsymbol{\omega} \wedge \boldsymbol{\rho})\, dm \\ &\quad + \int \boldsymbol{\rho} \wedge (\boldsymbol{\omega} \wedge \mathbf{r}_C)\, dm + \int \boldsymbol{\rho} \wedge (\boldsymbol{\omega} \wedge \boldsymbol{\rho})\, dm\end{aligned}$$

Noting that $\boldsymbol{\omega}$, $\dot{\mathbf{q}}$ and $\mathbf{r}_C$ can be taken outside the integrals, and that $\int \boldsymbol{\rho}\, dm = 0$, this becomes

$$\mathbf{H}_Q = (\mathbf{r}_C \wedge \dot{\mathbf{q}})m + \mathbf{r}_C \wedge (\boldsymbol{\omega} \wedge \mathbf{r}_C)m + \int \boldsymbol{\rho} \wedge (\boldsymbol{\omega} \wedge \boldsymbol{\rho})\, dm$$

The last term is $\mathbf{H}_C$ and the other two can be combined because

$$v_C = \dot{\mathbf{q}} + \boldsymbol{\omega} \wedge \mathbf{r}_C$$

So $$\mathbf{H}_Q = \mathbf{H}_C + \mathbf{r}_C \wedge m\, v_C \qquad (7.28)$$

We now have, in (7.24) and (7.28), expressions for the moment of momentum of a rigid body about (1) any point fixed in space and (2) any point fixed to the body.

Evaluation of H

To find $\mathbf{H}_O$ or $\mathbf{H}_Q$ we need an expression for $\mathbf{H}_C$, the moment of momentum of the body about its centre of mass.

(7.27) $$\mathbf{H}_C = \int \boldsymbol{\rho} \wedge (\boldsymbol{\omega} \wedge \boldsymbol{\rho})\, dm$$

The xyz components of $\mathbf{H}_C$ are given by

$$\mathbf{H}_C = H_x\hat{\mathbf{i}} + H_y\hat{\mathbf{j}} + H_z\hat{\mathbf{k}}$$

where $\hat{\mathbf{i}}, \hat{\mathbf{j}}, \hat{\mathbf{k}}$ are unit vectors in the xyz frame as in Chapter 5.
Writing

$$\boldsymbol{\rho} = x\hat{\mathbf{i}} + y\hat{\mathbf{j}} + z\hat{\mathbf{k}} \text{ and } \boldsymbol{\omega} = \omega_x\hat{\mathbf{i}} + \omega_y\hat{\mathbf{j}} + \omega_z\hat{\mathbf{k}}$$

the product in the integral can be evaluated, remembering that

$$\hat{\mathbf{i}} \wedge \hat{\mathbf{i}} = \hat{\mathbf{j}} \wedge \hat{\mathbf{j}} = \hat{\mathbf{k}} \wedge \hat{\mathbf{k}} = 0 \text{ and}$$
$$\hat{\mathbf{i}} \wedge \hat{\mathbf{j}} = -\hat{\mathbf{j}} \wedge \hat{\mathbf{i}} = \hat{\mathbf{k}};\ \hat{\mathbf{j}} \wedge \hat{\mathbf{k}} = -\hat{\mathbf{k}} \wedge \hat{\mathbf{j}} = \hat{\mathbf{i}};\ \hat{\mathbf{k}} \wedge \hat{\mathbf{i}} = -\hat{\mathbf{i}} \wedge \hat{\mathbf{k}} = \hat{\mathbf{j}}$$

$$\boldsymbol{\omega} \wedge \boldsymbol{\rho} = (z\omega_y - y\omega_z)\hat{\mathbf{i}} + (x\omega_z - z\omega_x)\hat{\mathbf{j}} + (y\omega_x - x\omega_y)\hat{\mathbf{k}}$$

$$\begin{aligned}\boldsymbol{\rho} \wedge (\boldsymbol{\omega} \wedge \boldsymbol{\rho}) = {} & [\omega_x(y^2 + z^2) - \omega_y xy - \omega_z xz]\hat{\mathbf{i}} \\ & + [-\omega_x yx + \omega_y(z^2 + x^2) - \omega_z yz]\hat{\mathbf{j}} \\ & + [-\omega_x zx - \omega_y zy + \omega_z(x^2 + y^2)]\hat{\mathbf{k}}\end{aligned}$$

Integration of the whole lot gives $\mathbf{H}_C$. Plainly therefore the three separate lines integrate to H_x, H_y and H_z. Writing $\int (y^2 + z^2)\, dm = I_{xx}$ and $\int xy\, dm = I_{xy}$, etc., they are

$$\left.\begin{aligned} H_x &= I_{xx}\omega_x - I_{xy}\omega_y - I_{xz}\omega_z \\ H_y &= -I_{yx}\omega_x + I_{yy}\omega_y - I_{yz}\omega_z \\ H_z &= -I_{zx}\omega_x - I_{zy}\omega_y + I_{zz}\omega_z \end{aligned}\right\} \qquad (7.29)$$

The full set of integrals for the body is

$$\begin{aligned} I_{xx} &= \int (y^2 + z^2)\, dm & I_{xy} &= \int xy\, dm \\ I_{yy} &= \int (z^2 + x^2)\, dm & I_{yz} &= \int yz\, dm \\ I_{zz} &= \int (x^2 + y^2)\, dm & I_{zx} &= \int zx\, dm \end{aligned} \qquad (7.30)$$

The three terms on the left are the *moments of inertia* about the x, y, z axes respectively, and we have met terms like these before. Note that the terms in the brackets are squares and so are always positive. The three terms on the right (only three because $I_{xy} = I_{yx}$, $I_{yz} = I_{zy}$ and $I_{zx} = I_{xz}$) are called the *products of inertia*, and we have not met them before. All six quantities describe the way in which the material in the body is distributed with respect to the xyz axes. They are true constants for a given set of xyz axes, because the xyz axes are fixed to the body. Of course their values will depend on *how* the xyz axes are attached to the body. You will see for instance that the terms in the product of inertia integrals can be positive or negative, and so it is possible for the integral to be zero. Indeed it can be shown that for any location of the origin of xyz coordinates, three perpendicular directions can be found for x, y, z such that the products of inertia are all zero. They are called *principal axes of inertia* and the moments of inertia measured about them are the *principal moments of inertia* for that particular point and body. Of course the directions of x, y, z and the values of I_{xx} I_{yy} and I_{zz} will depend, for any particular body, on the location of the x, y, z origin.

Returning to the evaluation of $\mathbf{H}$, and $\mathbf{H}_C$ in particular, the values of the I's in equations (7.29) are all with respect to the centre of mass, because (7.29) are the expressions for the components of $\mathbf{H}_C$, which is the moment of momentum about the centre of mass. We are now able, in principle at any rate, to evaluate the moment of momentum of a rigid body about any point fixed in space or any point fixed to the body.

Equations of Motion of a Rigid Body

The motion of the centre of mass of the body is determined by (7.21)

$$\Sigma\mathbf{F} = m\ddot{\mathbf{c}}$$

The rotation about the centre of mass is governed by the special case of (7.23) where P is located at the centre of mass,

$$\mathbf{M}_C = \dot{\mathbf{H}}_C$$

$\mathbf{H}_C$ is given by (7.29), but we must be careful when differentiating $\mathbf{H}_C$ to remember that the unit vectors $\hat{\mathbf{i}}, \hat{\mathbf{j}}, \hat{\mathbf{k}}$ vary in direction and so do have time derivatives. For xyz axes rotating with angular velocity $\boldsymbol{\omega}$

$$\dot{\hat{\mathbf{i}}} = \boldsymbol{\omega} \wedge \hat{\mathbf{i}}, \ \dot{\hat{\mathbf{j}}} = \boldsymbol{\omega} \wedge \hat{\mathbf{j}}, \ \dot{\hat{\mathbf{k}}} = \boldsymbol{\omega} \wedge \hat{\mathbf{k}}$$

$$\begin{aligned}\dot{\mathbf{H}}_C &= \frac{\mathrm{d}}{\mathrm{d}t}[H_x\hat{\mathbf{i}} + H_y\hat{\mathbf{j}} + H_z\hat{\mathbf{k}}] \\ &= \dot{H}_x\hat{\mathbf{i}} + H_x\dot{\hat{\mathbf{i}}} + \dot{H}_y\hat{\mathbf{j}} + H_y\dot{\hat{\mathbf{j}}} + \dot{H}_z\hat{\mathbf{k}} + H_z\dot{\hat{\mathbf{k}}}\end{aligned}$$

$$\begin{aligned}\dot{\hat{\mathbf{i}}} &= \boldsymbol{\omega} \wedge \hat{\mathbf{i}} = (\omega_x\hat{\mathbf{i}} + \omega_y\hat{\mathbf{j}} + \omega_z\hat{\mathbf{k}}) \wedge \hat{\mathbf{i}} \\ &= \omega_z\hat{\mathbf{j}} - \omega_y\hat{\mathbf{k}}\end{aligned}$$

Similarly,

$$\dot{\hat{\mathbf{j}}} = \omega_x \hat{\mathbf{k}} - \omega_z \hat{\mathbf{i}}$$
$$\dot{\hat{\mathbf{k}}} = \omega_y \hat{\mathbf{i}} - \omega_x \hat{\mathbf{j}}$$

This works out to

$$\dot{\mathbf{H}}_C = (\dot{H}_x - \omega_z H_y + \omega_y H_z)\hat{\mathbf{i}} + (\dots)\,\hat{\mathbf{j}} + (\dots)\,\hat{\mathbf{k}}$$

Now

$$\mathbf{M}_C = \dot{\mathbf{H}}_C$$

so we can write the three component equations

$$\begin{aligned} M_x &= \dot{H}_x - \omega_z H_y + \omega_y H_z \\ M_y &= \dot{H}_y - \omega_x H_z + \omega_z H_x \\ M_z &= \dot{H}_z - \omega_y H_x + \omega_x H_y \end{aligned} \qquad (7.31)$$

If we now substituted (7. 29) into (7. 31) we would obtain the general equations of rotational motion. They would be very complicated, and we shall make the substitution only for a few special cases. These will be plane motion (to confirm assertions made earlier in the chapter), and motion with the origin of the *xyz* axes fixed in space (but not necessarily at the centre of mass).

Plane motion

Let the *xy* and *XY* planes coincide and be the plane of the motion. The *xy* axes are fixed to the body with the origin at the centre of mass.

$$\ddot{z} = \omega_x = \omega_y = 0$$

For the centre of mass,

$$F_x = m\ddot{x}_C : F_y = m\ddot{y}_C : F_z = 0$$

(7. 29) reduce to

$$\begin{aligned} H_x &= -I_{xz}\,\omega_z \\ H_y &= -I_{yz}\,\omega_z \\ H_z &= I_{zz}\,\omega_z \end{aligned}$$

Substituting these into (7. 31) gives

$$\begin{aligned} M_x &= -I_{xz}\,\dot{\omega}_z + I_{yz}\,\omega^2{}_z \\ M_y &= -I_{yz}\,\dot{\omega}_z - I_{xz}\omega^2{}_z \\ M_z &= I_{zz}\,\dot{\omega}_z \end{aligned} \qquad (7.32)$$

These expressions include those given earlier for plane motion, (7. 17) and (7. 18) where $I_C = I_{zz}$, but we now have the additional terms M_x and M_y. We see in (7.32) that, to maintain plane motion, there must be moments about the x and y axes of the body unless I_{xy} and I_{yz} are both zero. Think about connecting rods and other machine parts.

The moment of momentum about a point fixed in either the XY plane or the xy plane (which is moving with the body) is, from (7.28)

$$\mathbf{H}_Q = \mathbf{H}_C + \mathbf{r}_C \wedge m v_C$$

which confirms the expression (7.19).

Rotation about a fixed point

For this special case we fix the origin of xyz coordinates at the XYZ origin.

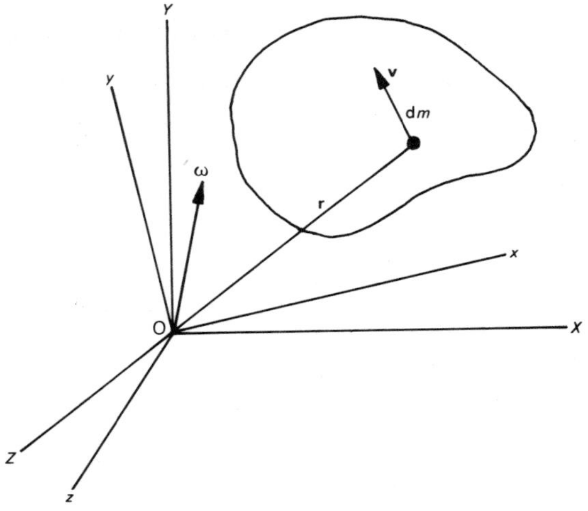

FIG. 7.21

In Fig. 7.21 the body is fixed in the xyz axes, which rotate with angular velocity $\boldsymbol{\omega}$ about their common origin O with the XYZ frame. The moment of momentum of the element dm about O is

$$\mathbf{r} \wedge dm\, v$$

The moment of momentum about O of the whole body is

$$\mathbf{H}_O = \int \mathbf{r} \wedge v\, dm$$

$$v = \boldsymbol{\omega} \wedge \mathbf{r} \qquad (\dot{\mathbf{R}} = 0 \text{ in } (7.26))$$

so

$$\mathbf{H}_O = \int \mathbf{r} \wedge (\boldsymbol{\omega} \wedge \mathbf{r})\, dm \qquad (7.33)$$

This has exactly the same form as (7. 27), but its meaning is not quite the same. (7. 27) gives the value of $\mathbf{H}_C$, the moment of momentum about the moving centre of mass. (7. 33) gives the value of $\mathbf{H}_O$, the moment of momentum about the fixed point which is the centre of rotation of the body, not in general its centre of mass.

The evaluation of $\mathbf{H}_O$ will follow the same lines as that of $\mathbf{H}_C$, and the expressions will be (7. 29) again, but now the I's will be measured with respect to O, the centre of rotation, *not* the centre of mass.

$$\begin{aligned} \mathbf{H}_O &= H_x\hat{\mathbf{i}} + H_y\hat{\mathbf{j}} + H_z\hat{\mathbf{k}} \\ H_x &= I_{xx}\,\omega_x - I_{xy}\,\omega_y - I_{xz}\,\omega_z \\ H_y &= -I_{yx}\,\omega_x + I_{yy}\,\omega_y - I_{yz}\,\omega_z \\ H_z &= -I_{zx}\,\omega_x - I_{zy}\,\omega_y + I_{zz}\,\omega_z \end{aligned} \qquad (7.34)$$

where the I's are now measured about xyz axes with origin at the centre of rotation.

We can now look again at the case of *Rotation about a fixed axis*, the case we considered in a simple way in Fig. 7. 3. The z and Z axes coincide and are the axis of rotation.

$$\begin{aligned} \omega_x &= \omega_y = 0 \\ H_x &= -I_{xz}\,\omega_z \\ H_y &= -I_{yz}\,\omega_z \\ H_z &= I_{zz}\,\omega_z \end{aligned}$$

Putting these into (7. 3) gives the equations of motion

$$\begin{aligned} M_x &= -I_{xz}\,\dot{\omega}_z + I_{yz}\,\omega^2{}_z \\ M_y &= -I_{yz}\,\dot{\omega}_z - I_{xz}\,\omega^2{}_z \\ M_z &= I_{zz}\,\dot{\omega}_z \end{aligned} \qquad (7.35)$$

The last of these is the same as (7. 7), and gives the moment required to drive the rotation. But now in addition we have the two other moments, M_x and M_y, which are required to maintain the motion about the z axis, even if the motion is steady, i.e. $\dot{\omega}_z = 0$. Remembering that the xyz axes are fixed to the body, M_x and M_y will also rotate with the body, causing vibration of the assembly.

To make M_x and M_y zero, it is necessary to make I_{xz} and I_{yz} both zero, in other words to make the axis of rotation a principal axis of inertia. Engineers go to a lot of trouble to make the products of inertia equal to zero in high speed rotating machinery—that is what rotor balancing is all about.

Returning to the general case of *Rotation about a fixed point*, the equations governing the motion are (7. 21), (7. 31) and (7. 34). The combination of (7. 31) and (7. 34) would be immensely complicated without some simplification. A major step in this direction is the restriction of the analysis to bodies with an *axis of circular symmetry*, i.e. bodies of revolution. If the axis of symmetry is the z axis, then

$$I_{xx} = I_{yy} = I, \text{ say}, \qquad I_{xy} = I_{yz} = I_{zx} = 0, \text{ and}$$

I_{zz} will be written simply I_z.

Once this step has been taken, there is no longer any need to rotate the xyz axes with the body, because $I_{xx} = I_{yy} = I$ will still hold so long as the body's axis of symmetry coincides with the z axis. In other words it is in order to allow the body to *spin about* the z axis with angular velocity Ω, provided we modify the expression for **H** accordingly:

$$H_x = I\,\omega_x,\ H_y = I\,\omega_y,\ H_z = I_z\,(\omega_z + \Omega) \tag{7.36}$$

are now the modified version of (7. 34).

Substituting (7. 36) into (7. 31) gives the equations of motion,

$$\begin{aligned} M_x &= I\,(\dot{\omega}_x - \omega_z\omega_y) + I_z\,(\omega_z + \Omega)\omega_y \\ M_y &= I\,(\dot{\omega}_y + \omega_x\omega_z) - I_z\,(\omega_z + \Omega)\omega_x \\ M_z &= I_z\,(\dot{\omega}_z + \Omega) \end{aligned} \tag{7.37}$$

You will note in the first two equations that the moments arising from the spin of the body about its own axis, Ω, are a moment about the x *axis involving* ω_y, and a moment about the y *axis involving* ω_x. These are called *gyroscopic moments*. It is the angular velocity produced at right angles to the external moment which is responsible for the apparently incongruous behaviour of the gyroscope. Anyone who has played with a toy gyroscope or a bicycle wheel will be familiar with the phenomenon. Anyone who has not should try the simple experiment with a bicycle wheel described below. But first we complete our brief flirtation with three-dimensional dynamics by looking at the classical coordinates used to describe motion about a fixed point.

Now that the body can spin about the z axis, it is much easier to describe its motion by restricting one of the other axes, the x axis in Fig. 7. 22, to move in the XZ plane. This involves no restriction on the motion of the body. The position of the xyz axes is then defined by the two angles θ and ϕ, and the motion of the body relative to the xyz axes by Ω about the z axis. The yz plane always contains the Y axis.

$\boldsymbol{\omega}$ is the angular velocity of the xyz frame, so its components are

$$\begin{aligned} \omega_x &= \dot{\theta} \\ \omega_y &= \dot{\phi}\sin\theta \\ \omega_z &= \dot{\phi}\cos\theta \end{aligned} \tag{7.38}$$

Substituting (7. 38) into (7. 37) would give the governing equations of the motion of a symmetrical body about a fixed point. They are complicated, and no general solution has been obtained, despite much attention over many years. The motion of a spinning top is a classical problem which has attracted much attention

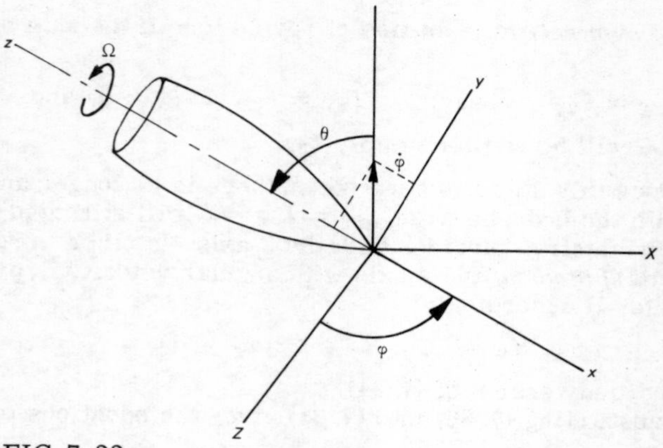

FIG. 7. 22

in its own right, but a closely related practical problem is that of the flight of a spin-stabilised artillery shell. The rifling in the gun gives the shell a fast spin about its axis, and the gyroscopic couple overcomes the tendency for the shell to 'topple' under the influence of the aerodynamic 'overturning moment'.

One of the many special cases for which solutions have been found, and much the simplest, is the steady precession of a gyroscope. You can easily demonstrate this for yourself by hanging a bicycle wheel from the ceiling by a piece of string attached to one end of the spindle. If you let the wheel hang at rest it will of course settle in a near-horizontal position, but if you hold the wheel with the spindle horizontal, and spin the wheel before releasing it, the spindle will stay horizontal when you let go, but it will slowly rotate about the vertical—the motion called *precession*.

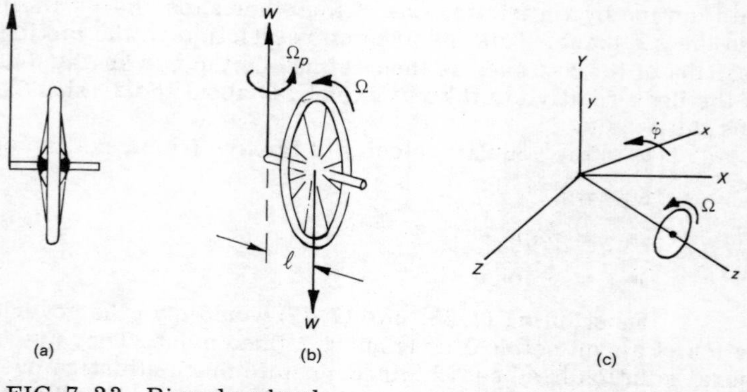

FIG. 7. 23 Bicycle wheel

The set-up is shown in Fig. 7.23, where the rate of steady precession is indicated by Ω_p. Comparing Figs. 7.22 and 7.23, we see that $\theta = \frac{\pi}{2}$, $\dot{\theta} = 0$ and $\dot{\phi} = \Omega_p$. Expressions (7.38) become

$$\omega_y = \Omega_p, \quad \omega_x = \omega_z = 0$$

Into (7.37)

$$M_x = I_z\Omega\Omega_p, M_y = I\dot{\Omega}_p, M_z = 0$$

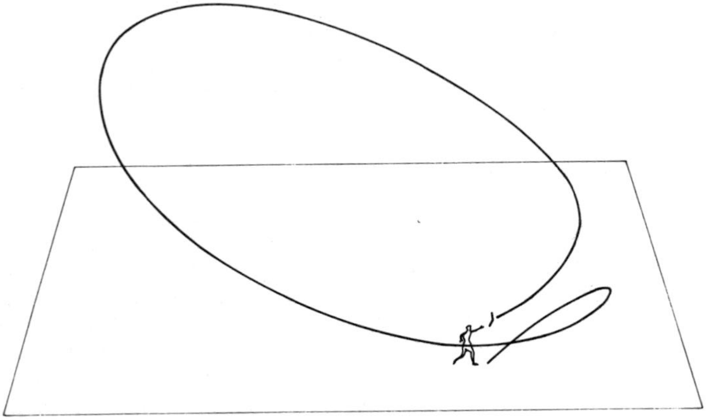

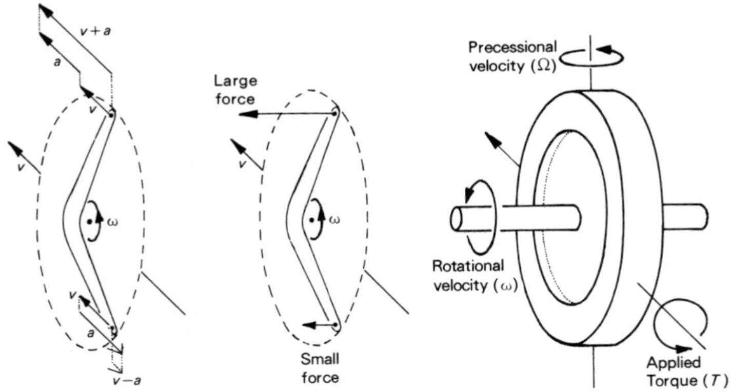

FIG. 7.24 The aerodynamics of boomerangs

The only applied moment is that due to the weight of the wheel; it is Wl and is clockwise when viewed along the x axis, i.e. it is positive.

$$M_x = Wl = I_z\Omega\Omega_p$$

$$M_y = 0 = I\dot{\Omega}_p, \ \Omega_p = \text{constant}$$

So
$$\Omega_p = \frac{Wl}{I_z\Omega} \tag{7.39}$$

The faster the wheel spins, the slower the precession.

This simple experiment demonstrates the characteristic of the gyroscope: on three mutually perpendicular axes, the axial spin, the external moment and the precession.

Many applications of the gyroscope, including the gyrocompass, stabilisers and inertial navigation, are described and analysed in [3]. A less technological, but no less interesting, application is described in a delightful article [4] from which Fig. 7. 24 is taken.

Elastic bodies

In Chapter 4 on statics we asked 'How stiff is rigid?' in connection with the subject of rigid body statics. We round off this introductory text to engineering mechanics by asking the same question about rigid body dynamics. Obviously, part of the answer is the same as for statics: if the body is to be treated as rigid it must be *capable* of transmitting the applied forces without breaking or so deforming as to change its geometry significantly. But there is an additional consideration in the dynamic situation—that of *time*. In statics we were concerned with steady or slowly changing patterns of applied force. In dynamics things may be changing very quickly: in a racing car engine the connecting rods go through their cycle of loading hundreds of times in a second; we do not know how long the collision between two billiard balls lasts, but we are sure that it is not very long.

A question we must ask therefore is 'When a force is applied to a body, how long does it take to be transmitted through the body?' The full answer is complicated, but we can get an idea of the nature of the phenomenon by imagining what goes on when a long, thin bar of steel is dropped vertically on to a large steel block on the floor. Figure 7. 25 shows roughly what happens.

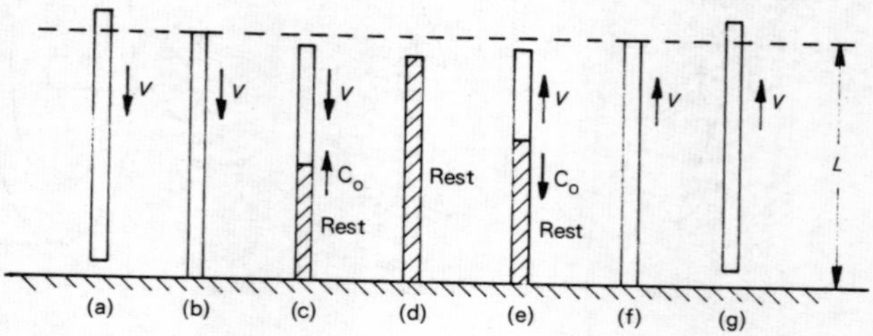

FIG. 7. 25 Bouncing bar

In Fig. 7.25 (a) the bar is falling vertically at velocity V, and is about to strike the floor, which we shall assume to be *rigid* for this illustration. At (b) the bottom end of the bar reaches the floor and is therefore brought to rest. How does the top end of the bar know that the bottom has struck the floor?—it doesn't. The information is conveyed upward by a compressive stress wave which travels up the bar at velocity, say, C_0 as in (c); below the wave front the material of the bar is at rest and compressed, above the wave front it is still stress-free and moving downwards with velocity V. Obviously the wave must travel very fast if the whole bar is not to be concertinered like a character in a cartoon. It can be shown* that the wave velocity C_0 for this simple case is $\sqrt{E/\rho}$, where E is Young's modulus and ρ is the density of the bar.

When the wave reaches the top of the bar (d), the whole bar is compressed and instantaneously at rest. The compression wave is reflected from the free surface as an equal tension wave, and travels down the bar releasing the compressive stress. For a perfectly elastic bar the whole process is reversed in (e) and (f), and the bar rebounds in (g) with velocity V.

If the length of the bar is L, the time taken for the whole bounce will be $2L/C_0$. This will be the periodic time of the lowest in an infinite number of longitudinal vibration frequencies, and will therefore be the *longest* of those possible.

For steel, the relevant figures are

$$E = 2 \times 10^{11}\ \mathrm{N/m^2}, \quad \rho = 7.8 \times 10^3\ \mathrm{kg/m^3}$$

giving

$$C_0 = \sqrt{\frac{E}{\rho}} \approx 5\,000\ \mathrm{m/s}$$

The order of time involved in this sort of wave propogation will be

$$l\sqrt{\frac{\rho}{E}}$$

where l is the length of the component in the machine or whatever. For high-speed oscillating machine components we are probably not far out to say

$$l \sim 0.1\ \mathrm{m\ or\ less}$$

so

$$l\sqrt{\frac{\rho}{E}} \sim 10^{-5}\ \mathrm{s}$$

So we would expect no threat to our assumption of rigidity on this score, when we compare 10^{-5} s with 'hundreds of times in a second' in a racing car engine.

The period of longitudinal vibrations appears to be too short to present a problem in any except most unusual circumstances. What about transverse (bending) vibrations? The frequencies depend on the end conditions, but the period of the fundamental mode (the slowest) is of the order

$$l\sqrt{\frac{\rho}{E}} \times \frac{l}{k}$$

* See [5], for example.

where k is a transverse radius of gyration. This period is l/k times that of longitudinal vibrations. Now l/k will usually be 10 or more so the characteristic time will be much longer than the 10^{-5} s above. If in addition the link we are thinking of is connected to some much larger mass than its own, the link becomes merely the spring in the system, and the period of vibration may be increased vastly. Vibrations in that class most certainly are a design consideration, and are the subject of a book by McCallion [6], in the related series of Advanced Engineeri Texts.

In this long last chapter we have studied the dynamics of bodies, first in a simple way which will suffice for most branches of engineering; and then in a more detailed way which will serve as an introduction to those who intend to pursue the subject.

Bibliography

1. Morrison, J. L. M. and Crossland, B. *An Introduction to the Mechanics of Machines,* Longmans, 1964
2. Rutherford, D. E. *Classical Mechanics,* Oliver and Boyd, 1957
3. Arnold, R. N. and Maunder, L. *Gyrodynamics and its Engineering Applications* Academic Press, 1961
4. Hess, F. *The aerodynamics of boomerangs,* Sci. Amer, Nov. 1968
5. Kolsky, H. *Stress Waves in Solids,* Oxford, 1953
6. McCallion, H. *Vibration of Linear Mechanical Systems,* Longmans, 1972
7. Housner, G. W. and Hudson, D. E. *Dynamics,* Van Nostrand, 1959
8. Swanson, S. A. V. *Engineering Dynamics,* English Universities Press, 1963

Examples

$g = 9.81$ m/s^2

Ex. 7.1 A motor car with rear wheel drive has a wheelbase length of 3 m. The centre of mass is 1 m rear of the front wheel axis and 0.6 m up from the ground.

On the assumption that there is ample power available from the engine, calculate

(a) the maximum acceleration attainable with a coefficient of friction between tyres and ground of 0.6, and
(b) the lowest coefficient of friction, and the acceleration, at which the front wheels will just leave the ground.

Neglect all rotational inertias. [2.2 m/s^2, 3.33, 33 m/s^2]

Ex. 7.2 A uniform beam rests on two simple supports at the ends. If one of the supports is suddenly removed, what will be the instantaneous reaction at the other support? [$W/4$]

Ex. 7.3 A uniform beam of mass 10 kg and length 1 m is pivoted at one end and supported on a linear spring at the other. When the beam is pressed against the spring into the horizontal position, as shown in the figure, the force in the spring

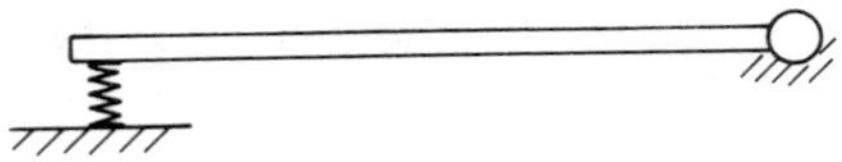

Ex. 7. 3

is 5 000 N and the compression in the spring is 100 mm. If the beam is released from that position, what will be its angular velocity and the force at the pivot when it passes through the vertical position. [11 rad/s, 505 N]

Ex. 7. 4 For test running a small rocket motor is connected to a central pivot by a long slender tie-bar. When running the motor moves in a circular path in a horizontal plane. The tie-bar has mass m per unit length and its length between the pivot and the motor is a.

Derive an expression for the bending moment distribution along the tie-bar when the motor is running up to speed and giving the whole assembly an angular acceleration $\ddot{\theta}$. Ignore the bending moment due to gravity.

$$\left[\frac{m\ddot{\theta}}{6}r(a^2 - r^2)\right]$$

Ex. 7. 5 An electric motor, whose rotating parts have a moment of inertia I_1, drives a rotor of inertia I_2 through a variable gear ratio. If the motor torque is constant, what gear ratio (driven: driver) should be selected to give the rotor the maximum angular acceleration? $[\sqrt{I_1/I_2}]$

Ex. 7. 6 A motor car of total mass 1 000 kg has wheels 0. 75 m diameter and total moment of inertia 8 kg m^2. The engine shaft has a moment of inertia of 0. 5 kg m^2 and is geared to the driving wheels through a 6 to 1 reduction. The engine exerts a torque of 200 N m and the friction torque on the engine shaft is 20 N m. The total friction torque on the front and back axles is also 20 N m, and the other resistances to motion (air resistance, tyre rolling resistance) are equivalent to a force of 250 N on the car itself.

Estimate the acceleration of the car on a level road. [2. 2 m/s^2]

Ex. 7. 7 Two uniform discs are freely pivoted about vertical axes as shown in the figure. A small electric motor, mounted on disc B, is directly coupled to disc A, and has a running speed of 200 rev/min.

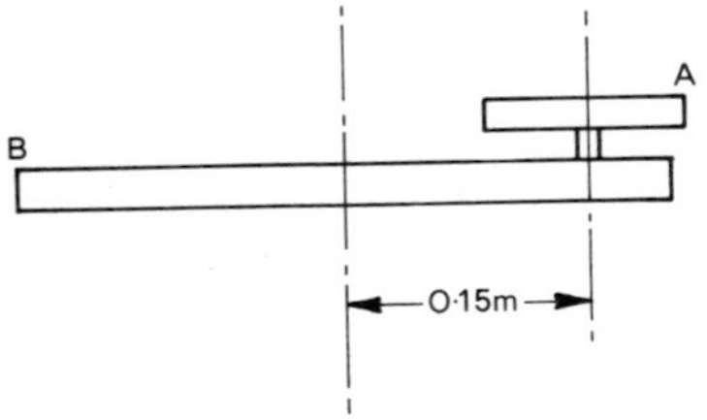

Ex. 7. 7

The masses of A and B are 10 kg and 20 kg, and their polar moments of inertia are 0. 05 kg m^2 and 0. 4 kg m^2 and the distance between their axes is 0. 15 m

If the motor is started when the system is at rest, estimate the absolute angular velocities of the discs when the motor attains its running speed. Neglect the mass of the motor. (No *external* moments, so total moment of momentum about axis of B is zero).

[185, -15 rev/min]

Ex. 7. 8 For the engine mechanism described in Example 5. 2, find the magnitude and direction of the torque on the crank resulting from the inertia of the connecting rod, which has a mass of 10 kg and a moment of inertia of 1 kg m^2 about its centre of mass, which is 0. 35 m from P.

[11 N m anticlockwise]

Ex. 7. 9 An experimental car is fitted with a gyro-stabiliser, mounted in fixed bearings on a shaft parallel to the rear axle, to counteract the tendency of the car to roll over when going round a bend.

The gyro rotor has mass m and radius of gyration k. The total mass of the car is M and the height of the centre of mass above the road is h. The speed of the car is v.

Derive an expression for the angular velocity of the gyro rotor, if the tendency to roll over is to be exactly counteracted. What will be the direction of spin for left and right-hand curves?

[Mvh/mk^2]

Appendix: Moments of Inertia

Thin rod

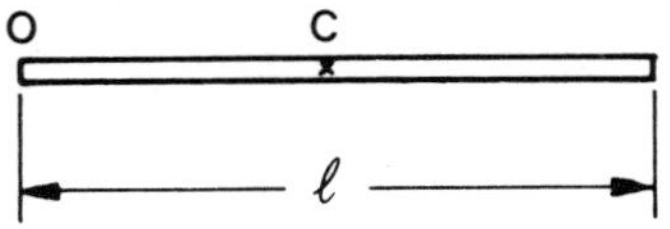

$$I_C = \frac{m\ell^2}{12}$$

$$I_O = \frac{m\ell^2}{3}$$

Thick beam

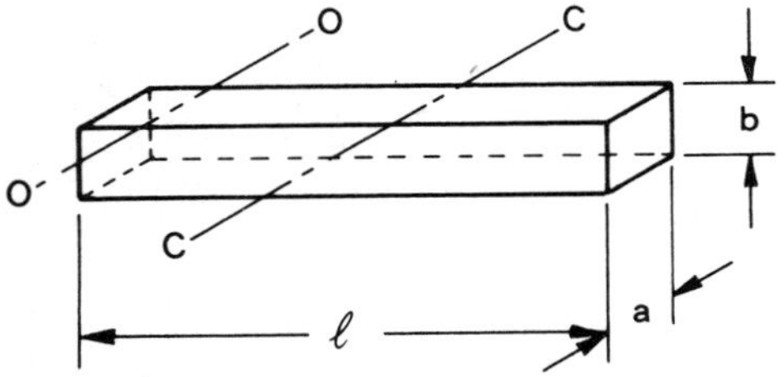

$$I_{CC} = \frac{m}{12}(\ell^2 + b^2)$$

$$I_{OO} = \frac{m}{12}(4\ell^2 + b^2)$$

Cylinder

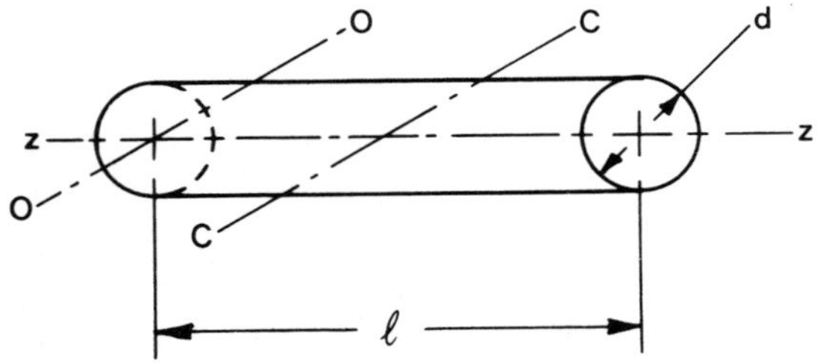

$$I_{zz} = \frac{md^2}{8}$$

$$I_{CC} = \frac{m}{12}\left(\ell^2 + \frac{3d^2}{4}\right)$$

$$I_{OO} = \frac{m}{12}\left(4\ell^2 + \frac{3d^2}{4}\right)$$

Sphere

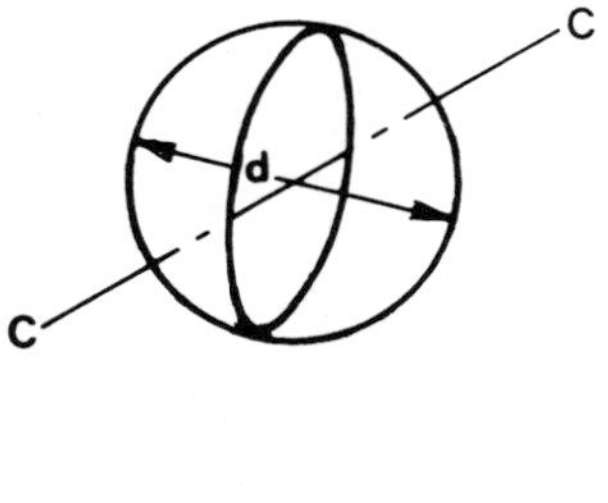

$$I_{CC} = \frac{md^2}{10}$$

Index